江西理工大学优秀博士论文文库

三苯基甲烷染料的微生物脱色

潘涛 著

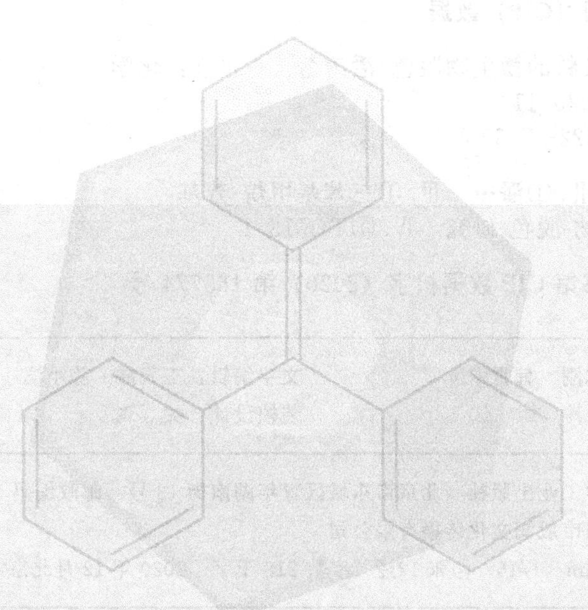

化学工业出版社

·北京·

内 容 简 介

本书以三苯基甲烷染料的来源、脱色方法和机理为主线讨论了其微生物脱色原理。首先介绍了三苯基甲烷染料的基础知识及其危害，并综述了三苯基甲烷染料的脱色方法和影响因素，在此基础上，详细分析了浊点系统中嗜水气单胞菌对三苯基甲烷染料的萃取微生物脱色，包括菌种的脱色机理、染料在浊点系统中的分配规律、浊点系统中的萃取微生物脱色，以及增溶剂的回收与再利用。

本书可供从事微生物降解研究的相关人员及污水处理相关的技术人员参考使用。

图书在版编目（CIP）数据

三苯基甲烷染料的微生物脱色/潘涛著. —北京：化学工业出版社，2020.11
ISBN 978-7-122-37660-2

Ⅰ.①三… Ⅱ.①潘… Ⅲ.①三苯基甲烷-芳基甲烷染料-微生物-脱色-研究 Ⅳ.①TQ613.1

中国版本图书馆 CIP 数据核字（2020）第 165774 号

责任编辑：仇志刚　韩霄翠　　　文字编辑：王云霞　陈小滔
责任校对：边　涛　　　装帧设计：关　飞

出版发行：化学工业出版社（北京市东城区青年湖南街 13 号　邮政编码 100011）
印　　装：涿州市般润文化传播有限公司
710mm×1000mm　1/16　印张 12½　字数 218 千字　2020 年 12 月北京第 1 版第 1 次印刷

购书咨询：010-64518888　　　售后服务：010-64518899
网　　址：http://www.cip.com.cn
凡购买本书，如有缺损质量问题，本社销售中心负责调换。

定　　价：88.00 元

前言

合成染料是一类重要的有机化合物，广泛应用于纺织、造纸、胶片、皮革、塑料、食品、化妆品和医药行业。目前，我国合成染料和颜料行业的年产量已超过100万吨。在生产和使用过程中，每年约有11%以上的染料会随废水排放到环境中。在水环境中，少量染料便可阻碍光线进入深水系统，导致藻类和浮游植物的光合作用受到干扰甚至被抑制。其中，三苯基甲烷染料结构稳定，难以被生物降解，造成了大量环境问题，引起广泛关注。

本书试图以三苯基甲烷染料的来源、脱色方法和机理为主线讨论其微生物脱色原理。本书第1章介绍了染料的历史、发展和组成，并详细阐述了三苯基甲烷染料及其危害；第2章综述了三苯基甲烷染料的脱色方法和影响因素，重点讨论了微生物吸附和微生物脱色以及脱色酶；第3章到第6章详细分析了浊点系统中嗜水气单胞菌对三苯基甲烷染料的萃取微生物脱色。其中，第3章介绍了菌种的脱色机理，第4章介绍了染料在浊点系统中的分配规律，第5章分析了浊点系统中的萃取微生物脱色，第6章主要涉及增溶剂的回收与再利用。本书是笔者对博士期间的研究成果和国内外相关文献的全面综述，力争描绘出三苯基甲烷染料萃取微生物脱色的研究前景。限于笔者的知识和见解，疏漏之处恳请读者不吝指正。

在本书出版之际，十分感谢我的博士生导师广东省微生物研究所的郭俊研究员的谆谆教诲，感谢广东省微生物研究所的孙国萍和许玫英研究员的悉心指导。笔者在萃取微生物降解领域的一点见解和收获得益于上海交通大学王志龙研究员的倾囊相授和华南理工大学吴振强教授的多年教导。书中涉及浊点系统中有机污染物的微生物降解的相关研究内容得到了国家自然科学基金（NO. 21866015，21407070）、江西省自然科学基金（NO. 20192BAB203016，20151BAB213019）和华南应用微生物国家重点实验室开放课题（NO. sklam002—2015）的资助。江西理工大学对本书出版提供经费资助。特此声明并致以诚挚的感谢。

潘涛

江西理工大学

2020年4月

目录

第4章 三苯基甲烷染料在浊点系统中的分配 / 97

第5章 浊点系统中三苯基甲烷染料的微生物脱色 / 133

第1章
三苯基甲烷染料及其危害

1.1 染料的历史及发展

1.1.1 染料的历史

人类对个性化空间和服装的需求可以追溯到几千年前。在这种需求中，颜色起着非常重要的作用。无论是这个星球上的矿物还是动植物，都能成为我们所使用的天然染料的来源。在古希腊和古罗马时代，衣服颜色是人们地位的象征。法律上禁止除精英以外的任何人穿紫色衣服，这种颜色因其价格昂贵和难以从贝类中获得而备受推崇（图 1-1）[1]。根据后来古罗马人的记载，一只海螺只能贡献一滴原液。染料制作耗费巨大，过程烦琐，大量推罗人投身其中。作为回报，到公元 3 世纪，推罗紫染色的羊毛已经与等重的黄金等价。中世纪新颜料的主要推动力是艺术的发展。逐渐地，在光或某些大气条件下容易褪色的颜料被更稳定的产品所取代。

(a)

(b)

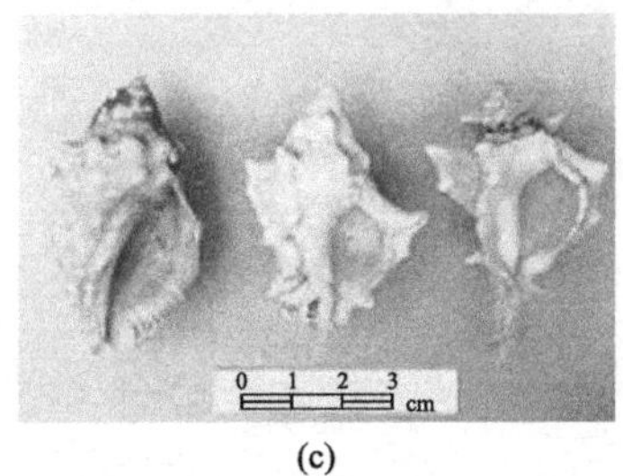

(c)

图 1-1　古希腊和古罗马时代的推罗紫[1]

(a) 颜色展示；(b) 结构式；(c) 用于提取推罗紫的贝类

染料的巨大变化发生在苯胺紫（mauveine）被发现之后。1856 年，在英国皇家化学学院著名有机化学家霍夫曼（Hoffmann）院长的实验室里，18 岁的研究生 W. H. Perkin 正在进行合成抗疟疾特效药物奎宁（quinine）的工作，当时这种药物必须从南美印第安人居住地的一种金鸡纳树的树皮中提取，因此该药物在欧洲的价格十分昂贵。Perkin 发现这种无法在棉布上染色的物质，却可以非常容易地染在丝绸和毛料上，而且比当时各种植物染料的颜色都鲜艳，放在肥皂水中搓洗也不褪色。这就是世界上第一种人工合成的化学染料苯胺紫（图 1-2）。Perkin 虽然没有合成出奎宁来，却获得了合成苯胺紫的发明专利。尽管苯胺紫仅仅保持了几年的商业地位，但它带动了一个巨大的新产业的发展。特别是在欧洲，这种统治一直持续到 20 世纪末。在 20 世纪，两种非常重要的人工合成染料发展起来了，一是酞菁（phthalocyanines），二是活性染料（reactive dyes）。它们令纤维素纤维的染色达到以前只有通过非常复杂和昂贵的工艺才能达到的牢度水平。

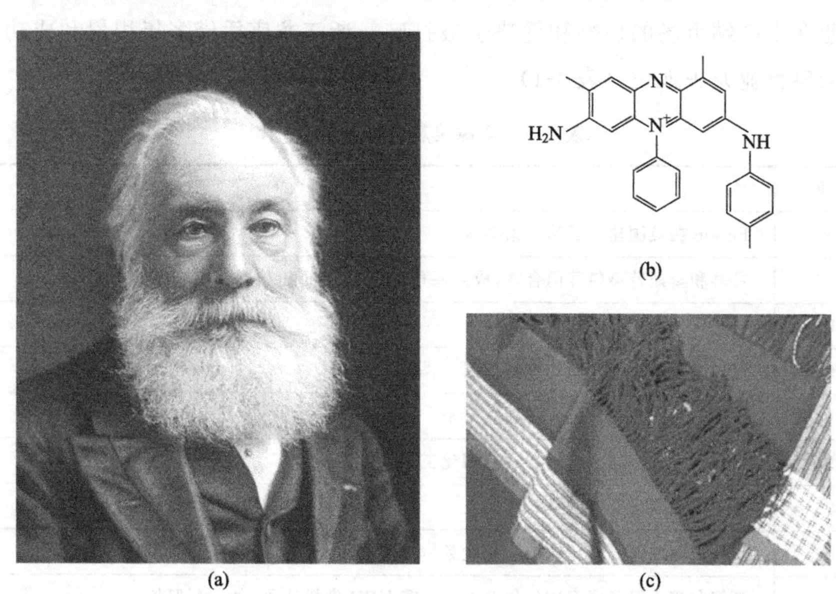

图 1-2　William Henry Perkin 和苯胺紫

(a) Perkin[2]；(b) 苯胺紫结构式；(c) 苯胺紫染色的围巾[3]

1.1.2　染料行业发展概况

20 世纪 50 年代，Pattee 和 Stephen 发现含二氯均三嗪基团的染料可在碱性条件下与纤维上的羟基发生键合，标志着染料使纤维着色从物理过程发展到化学过程，开创了活性染料的合成应用时期[4]。

20 世纪初期，德国染料工业发展突飞猛进，几乎实现了全球垄断。但由于劳动力成本及制造期间的高污染特性，使其不容于环保政策严厉的发达国家。因此，在 20 世纪末期，染料生产逐渐转向亚洲国家。巴斯夫（BASF）、拜耳（Bayer）、赫斯特（Hoechst）等企业逐渐退出染料行业，并在 21 世纪初经过并购重组形成德司达（Dystar）、亨斯迈（Huntsman）和科莱恩（Clariant）三大跨国染料企业。

由于大部分印染技术较为成熟，进入门槛较低，中国和印度的染料产业迅速崛起，大批规模不一的企业涉足染料及其中间体的生产销售。因此，为了应对市场份额重新分布的冲击，发达国家的传统染料企业积极转型高端染料制造，以中、印企业的廉价半成品染料为原料进行加工，生产高端产品。经过2008年金融危机的冲击，Clariant和Huntsman裁员，Dystar破产并被浙江龙盛收购，欧美企业在中高端市场的份额和优势也被打破。浙江龙盛凭借多年积累和成功收购跃居染料行业龙头地位（表1-1）。

表1-1 全球染料发展大事件[5]

时间	事件
1857	Perkin在英国建立了第一家染料厂
1995	拜耳和赫斯特染料部门合并，成立德司达公司；瑞士山德士公司分出科莱恩公司
1995	科莱恩从山德士公司分拆出来
1996	巴斯夫收购了捷利康的染料部门
1997	汽巴精化从瑞士汽巴公司独立出来
1997	科莱恩合并了德国赫斯特的专用化工部门
2000	德司达收购巴斯夫的染料部门
2006	亨斯迈收购汽巴精化纺织染化业务
2007	浙江龙盛入股印度KIRI，合资成立龙盛-KIRI染料公司，占60%股权
2009	德司达因资金链断裂，正式进入破产程序
2010	浙江龙盛通过全资子公司桦盛公司出资2200万欧元认购新加坡KIRI公司可转换债券
2010	新加坡KIRI公司完成对德司达集团资产的收购(剥离德国总部所有负债)
2011	闰土股份完成收购约克夏60%股权
2012	浙江龙盛可转换债券转股成功，成功控股德司达，股权占比62.43%
2013	科莱恩分出纺织化学品、造纸特种化学品和乳液三个业务单元重组为独立公司——昂高
2015	昂高完成收购巴斯夫纺织化学品业务

数据来源：财通证券研究所。

我国染料产业受下游纺织产业需求增大和出口增多推动，在近十年来取得了重大进展。目前我国是世界染料产量第一大国，约占世界染料总产量的70%以上。据中国染料工业协会统计，2011—2016年，我国染料产量复合增速为3.7%，2016年染料总产量达92.8万吨。

1.2 染料的分类

从来源上来看，染料可以分为天然染料和合成染料，大规模生产中使用的通常都是合成染料。染料是吸收波长在可见光范围内（400～700nm）的化合物[6,7]。染料分子中负责光吸收的主要结构是生色团，即具有共轭双键的离域电子系统。有机分子对紫外/可见辐射的吸收与分子轨道之间的电子跃迁有关。吸收辐射的能量由下式给出：

$$\Delta E = E_1 - E_0 = h\nu = hc/\lambda \tag{1-1}$$

式中，E_0 是与分子的基本态相对应的能量，J；E_1 是激发态能量，J；h 是普朗克常数，6.626×10^{-34}J·s；ν 是电磁辐射频率，Hz；c 是光速，3×10^8m/s；λ 是波长，nm。

电子离域程度越大，跃迁能量越低，波长越长。为了使电子离域，双键必须与单键交替。就合成染料而言，苯环或萘环也会促进离域化[8]。生色团通常含有氮、氧和硫等杂原子，并带有非成键电子。通过将这些电子结合到芳香环的离域系统中，电子云的能量被改变，吸收辐射的波长将向可见光范围移动，化合物将着色。在许多情况下，染料含有额外的助色团，它们是吸电子或给电子的取代基，通过改变电子系统的总能量来引起或增强生色团的颜色。最重要的助色团是：羟基及衍生物，—OH，—OR；氨基及衍生物，$—NH_2$，—NHR，$—NHR_2$；磺酸基，$—SO_3H$；羧基，—COOH；硫化物，—SR[6,7]。一些助色团还增强了染料对纤维（天然或合成纤维）的亲和力。天然纤维包括纤维素（棉和亚麻）或蛋白质（羊毛和丝绸）。合成纤维有黏胶纤维、醋酸纤维素、聚酰胺、聚酯和丙烯酸纤维等[9]。根据现有的发色团，常见的染料类别如表 1-2 所示。

表 1-2　基于官能团的染料分类[10]

染料分类	官能团	举例
硝基染料	$—N(=O)O$ 硝基	OH, NO_2, NO_2 酸性黄 24

续表

染料分类	官能团	举例
亚硝基染料	—N=O 亚硝基	快绿
偶氮染料	—N=N— 偶氮基	NaO_3S—C₆H₄—N=N—C₆H₄—$N(CH_3)_2$ 甲基橙
三苯基甲烷染料	三苯基	结晶紫
酞染料	邻苯二甲酸酐	酚酞
靛蓝染料	靛蓝	酸性蓝 71

续表

染料分类	官能团	举例
蒽醌染料	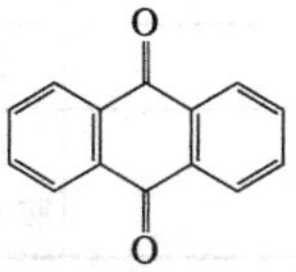 蒽醌	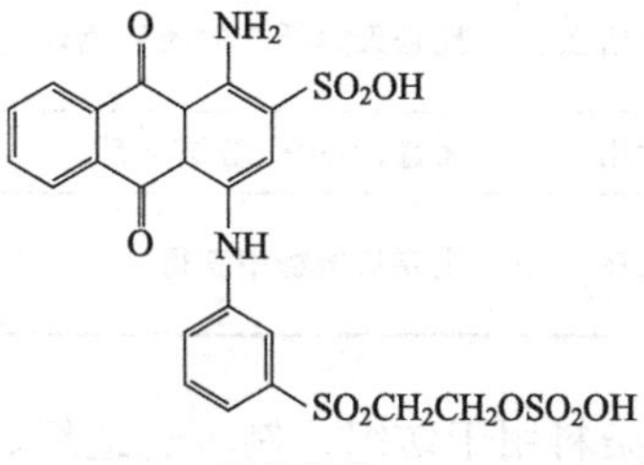 活性艳蓝 19

根据颜色指数（colour index，又称颜色索引），染料可按颜色和使用方法分类。各种力都可能使染料与纤维结合，相同的染料-纤维组合通常可以使用一种以上的化学键。主导力取决于纤维的化学特性和染料分子中的化学基团。在染料和纤维之间建立的键类型，相对强度从小到大的排序是：范德华力、氢键、离子键或共价键[8,9,11]。根据应用类别，染料可按表 1-3 分类。

表 1-3　基于应用类别的染料分类[10]

染料类型	特点	基质
酸性	当溶液中带负电时；与纤维中的阳离子—NH_3^+ 基团结合	羊毛、聚酰胺、丝绸、改性丙烯、纸张、油墨和皮革
活性	与—OH、—NH 或—SH 基团形成共价键	棉、毛、丝、聚酰胺
金属配合物	一种金属离子（通常是铬、铜、钴或镍）和一种或两种染料分子（酸性或活性）的强配合物	丝、毛、聚酰胺
直接	以范德华力与纤维结合的大分子	纤维素纤维、棉、黏胶、纸、皮革和聚酰胺
碱性	与纤维酸性基团结合的阳离子化合物	合成纤维、纸和油墨
媒染	需要添加与染料和纤维结合的化学物质，如鞣酸、明矾、铬明矾和其他铝、铬、铜、铁、钾和锡盐	羊毛、皮革、丝绸、纸张、改性纤维素纤维和阳极氧化铝
分散	难以溶解的染料，通过纤维膨胀渗透到纤维中	聚酯、聚酰胺、乙酸、丙烯酸
颜料	不溶性非离子化合物或不溶性盐，在应用过程中保持其晶体或颗粒结构	涂料、油墨、塑料和纺织品
还原	不溶性有色染料，还原后可形成对纤维具有亲和力的可溶无色形式；暴露在空气中会被重新氧化	纤维素纤维、棉、黏胶纤维和羊毛
偶氮和显色	纤维中偶合组分与重氮化芳香胺反应的不溶性产物	棉、黏胶、醋酸纤维素和聚酯
硫化	含 S 杂环的复合高分子芳香烃	纤维素纤维、棉和黏胶纤维
溶剂	非离子染料，可溶解与其结合的底物	塑料、汽油、清漆、涂料、污渍、墨水、油、蜡和脂肪

续表

染料类型	特点	基质
荧光增强	掩盖天然纤维的淡黄色调	肥皂和洗涤剂,所有纤维、油、涂料和塑料
食用	无毒,不用作纺织染料	食品
天然	主要从植物中获得	食品、棉花、羊毛、丝绸、聚酯、聚酰胺和聚丙烯腈

染料用于纺织、制革、造纸、食品、农业、激光阵列、光化学电池、染发剂和化妆品等行业。此外，染料还用于控制污水和废水处理的效力，确定活性污泥的比表面积和用于地下水追踪[12]。染料最重要的工业用途是纺织染色。

1.2.1 天然染料

天然染料是从自然资源（如植物、动物和矿物质）获得的，无需进行任何化学处理。来自靛蓝植物的靛蓝染料、散沫花的指甲花醌、Bignonia chica 的秋海棠色素和类胡萝卜素都是天然染料[13]。合成染料广泛应用于各个行业，但它们也是已知的诱变剂、过敏原和致癌物，而天然染料几乎没有毒性。从植物、动物或矿物质等自然资源中获得的天然染料非常清洁，因此非常环保[13]。天然染料除了可以染色外，还具有广泛的药用特性。如今，人们对天然染料的使用意识日益提高。由于天然染料几乎无毒、药用价值高、副作用小，目前已广泛应用于日常食品和医药行业[14,15]。

数百种植物可以用来提取天然染料。例如，由姜黄生产的一种天然亮黄色染料，具有很强的杀菌活性，可以治愈皮肤疾病[16]。天然染料几乎无毒，常用于食用色素、皮革、羊毛、丝绸、棉织物等天然蛋白质纤维，甚至在医药和化妆品等领域也广泛应用。由于合成染料的发展，世界范围内天然染料在纺织品中的使用仅限于工匠、小规模或家庭作坊级染坊。有些厂商也会生产一些用于纺织品的环保染料。天然染料需要一种称为媒染剂的化合物，用于将染料固定在织物上，防止染料渗出或被轻易洗掉。媒染剂有助于染料和纤维之间吸收染料的化学反应[16]。

大多数天然染料来自能产生各种颜色的植物。植物的许多部位，如种子、叶子、树皮、根、花、果实等，都会产生染料。有趣的是，迄今为止已经从不同植物中提取出的2000余种天然染料中，只有150种在市场上商业出售。迄今为止，很少有染料是从天然来源合成的。例如，紫草和碧霞珠等极少数植物，可以制造用于口红的靛红和眼影的靛蓝。染料在植物中的存在随植物的年龄和季节而变化[16]。许多植物，如石榴，由于含有大量的鞣酸而具有很高的抗菌潜力，可用于提取天然染料。此外，其他几种植物染料，如散沫花中提取的指甲花醌、核桃中提取的胡桃醌和紫草中提取的拉帕醌，都显示出抗菌特性[16]。天然染料的另一个优点是一些染料具有抗病原体的潜力。有报告指出五种天然染料（儿茶、紫胶虫、没食子、茜草和长刺酸模）对肺炎克雷伯菌、大肠杆菌、变形杆菌和铜绿假单胞菌具有抗菌活性[17]。其中，天然染料没食子的抗菌活性最高[17]。

1.2.2 合成染料

合成染料广泛应用于各种行业，如纺织品、皮革、美容产品、食品、医药和纸张印刷。造成环境污染的有毒化合物包括偶氮、蒽醌、杂环、三苯基甲烷和酞菁染料等[18]。印染废水在全球范围内引发多种环境问题。此外，废水的排放会污染地下水和地表水，这可能会导致人类和动物的各种健康问题，因为它们被认为是剧毒、致突变和致癌的。

(1) 偶氮染料

偶氮染料具有多种不同的形式和性质。目前，有超过 2000 多种性质迥异的偶氮染料在使用中。偶氮染料是合成染料中最大的一类[19]。工业上使用的染料大约 70%是偶氮染料。广泛应用于纺织、化妆品、皮革、医药、造纸、涂料、食品等行业。偶氮染料在当今染料化学中产量最大，在可预见的未来中会愈加重要。偶氮染料的巨大成功归功于以下几个因素：偶联反应的简便性，结构变化的巨大可能性以及多样化应用的适应性。据报道，全球每年约有 5 万吨纺织染料从染色过程中排放到环境中[19]。

偶氮染料结构多样，典型结构特征是存在偶氮键，即—N═N—。这种键型结构在偶氮分子中可能出现不止一次。单偶氮染料只有一个偶氮键，重氮染料中有两个偶氮键，而在三偶氮染料中有三个偶氮键。偶氮基两侧常与苯、萘和杂环芳烃相连[20]。侧链的不同决定了偶氮染料的颜色强度和色调[21]。

(2) 活性染料

活性染料主要用于染色纤维素纤维，如棉和黏胶，但它们对羊毛和聚酰胺也越来越重要。活性染料的应用范围很广，大量的染色技术可以采用。在用活性染料对纤维素纤维染色时，需使用以下化学品和助剂：

- 碱：碳酸钠，碳酸氢钠和氢氧化钠。

- 盐：主要是氯化钠和硫酸钠。
- 尿素：可以在连续工艺中添加到填充液中。
- 硅酸钠：可以通过冷轧分批法添加。

染料固定性差是活性染料长期存在的问题，特别是在纤维素纤维织物的成批染色中，通常会添加大量的盐来改善染料的上染率（因此也会提高染料的固定性）。因此，废水中的染料和盐是导致活性染料产生环境问题的主要因素。由于未固定活性染料及其水解产物均为水溶性的，因而在生物污水处理厂中难以去除。许多活性染料含有卤素。然而，由于活性染料不附着在生色团上，因此被认为不存在可吸附的有机卤素问题。

重金属既可以作为生产过程中的杂质，也可以作为生色团的组成部分。后者涉及酞菁染料，目前仍然广泛用于生产蓝色和青绿色色调。

(3) 硫化染料

硫化染料由高分子量化合物组成，通过硫或硫化物与胺和酚的反应获得。硫化染料是高度复杂的分子混合物，因此难以确切知晓其化学结构。硫化染料不溶于水，但是在碱性条件下被还原后，它们会转变为水溶性的无色形式，对纤维具有高亲和力。吸收到纤维中后，它们被氧化并转化为原始的不溶状态。

硫化染料主要用于棉和黏胶。它们也可用于染色合成纤维混合物中的纤维素，包括聚酰胺和聚酯。最受欢迎的硫化染料是硫黑，用来生产低成本黑色染料。硫化钠和硫氢化钠通常用作还原剂。在染色过程中，染料最终被氧化固定在基质上。过氧化氢或含卤素化合物，如溴酸盐、碘酸盐和亚氯酸盐，是最常用的氧化剂。除上述还原剂和氧化剂外，用硫化染料染色所需的其他化学品和助剂如下：

- 碱：主要是碳酸钠或氢氧化钠。
- 盐：氯化钠和硫酸钠。
- 分散剂：通常是萘磺酸-甲醛缩合物，木质素磺酸盐和磺化油。
- 络合剂：在某些情况下，使用乙二胺四乙酸（EDTA）和多磷酸盐可防止由于存在碱土金属离子而产生负面影响。

硫化染料氧化后不溶于水，通过污水处理厂的活性污泥吸附可大幅度去除。但存在生物降解性差的分散剂，而新的甲醛缩合产品具有更高的生物去除率（>70%）。

(4) 还原染料

从化学角度来看，还原染料可分为两类：靛蓝还原染料和蒽醌还原染料。与硫化染料一样，还原染料不溶于水，但在碱性条件下被还原后，它们变成水溶性物质浸透纤维。然后，它们通过氧化再次转化为原来不可溶的形式，并以这种形式固定在纤维中。还原染料最常用于棉、纤维素纤维及其混合物的染色。亚硫酸氢钠虽然有一定的局限性，但仍然是应用最广泛的还原剂。染色过程中常用的化学品和助剂如下：

- 亚硫酸氢钠、二氧化硫脲和亚砜酸衍生物作为还原剂；
- 氢氧化钠；
- 硫酸钠；
- 作为浸轧过程中抗迁移剂的聚丙烯酸酯和藻酸盐；
- 以萘磺酸和木质素磺酸盐作为分散剂的甲醛缩合产物；
- 表面活性剂（包括乙氧基化脂肪胺）和其他成分，如甜菜碱和聚烷基胺；
- 聚乙烯吡咯烷酮作为流平剂；
- 过氧化氢、过硼酸钠和 3-硝基苯磺酸作为氧化剂；
- 肥皂。

还原染料被氧化后是不溶于水的，因此它们可以通过吸附在污水处理厂的活性污泥上而被大量去除。还原染料含有生产过程中使用的重金属杂质（铜、铁、铅、钡、锰）。在某些情况下，将这些杂质控制在标准之下仍然很困难。

(5) 酸性染料

酸性染料基于偶氮、蒽醌、三苯基甲烷 Cu-酞菁生色体系，由于存在多达四个磺酸基团而可溶于水。这种染料主要用于聚酰胺和羊毛染色，还可用于丝绸和某些改性的丙烯酸纤维染色，但对纤维素和聚酯纤维几乎没有亲和力。酸性染料染色过程中最常用的化学品和助剂如下：

- 硫酸钠、乙酸钠和硫酸铵。
- pH 调节剂：乙酸、甲酸和硫酸，还有铵盐和磷酸盐。
- 匀染剂：主要是阳离子化合物，如乙氧基化脂肪胺。
- 后处理剂：如甲醛与芳香族磺酸缩合产物。

酸性染料是无毒的，但有两种染料——C. I. 酸性橙 150 和 165——已被分类为有毒染料。据报道，酸性紫罗兰 17 有致敏作用。酸性染料在废水中的存在率很低，是因为它们具有很高的上染率和固色度。

(6) 分散染料

分散染料的特点是低分子量和没有增溶基团。从化学角度看，超过 50% 的分散染料是简单的偶氮化合物，约 25% 是蒽醌，其余是次甲基、硝基或萘醌染料。分散染料主要用于聚酯，也用于乙酸纤维素和三乙酸纤维素、聚酰胺和丙烯酸纤维的染色。分散染料以粉末和液体产品形式提供。粉末染料包含40%～60%的分散剂，而在液体染料中，分散剂含量在10%～30%的范围内。甲醛缩合物和木质素磺酸盐被广泛用作分散剂。以下化学药品和助剂常用于分散染料的染色：

- 分散剂：尽管所有的分散染料都已经含有大量的分散剂，但染液和最后的洗涤步骤中仍会添加大量分散剂。
- 载体：对于聚酯纤维，在高达100℃的温度下用分散染料染色需要使用载体。由于与载体的使用相关的环境问题，聚酯优先在温度高于100℃且没有载体的情况下染色。然而，载体染色对于聚酯羊毛混纺织物仍然很重要。
- 增稠剂：在轧染过程中，通常将聚丙烯酸酯或藻酸盐加入到染液中。
- 还原剂（主要是亚硫酸氢钠）：在最后的洗涤步骤中与碱一起加入到溶液中，以除去未固定的表面染料。

由于分散染料的水溶性较差，通过污水处理厂的活性污泥吸附可大大消除分散染料。一些分散染料含有有机卤素，但由于其在活性污泥上的吸附作用，预计在处理后的废水中不会被发现。

(7) 碱性染料

碱性染料又称阳离子染料。碱性染料含有季胺基，是共轭体系的组成部分。碱性染料专门用于腈纶、改性聚酰胺和聚酯纤维以及混纺织物的染色。碱性染料通过静电引力与纤维紧密结合，不易迁移。为了实现匀染，通常使用特定的匀染助剂（称为缓凝剂）。最重要的一类助剂是具有长烷基侧链的季铵化合物（阳离子缓凝剂）。许多碱性染料表现出很高的水生毒性，但如果使用得当，它们显示出接近100%的染料活性。常见的环境问题通常是由处理程序不当、泄漏清理不彻底和其他干扰造成的。

(8) 直接染料

直接染料可以是偶氮化合物、二苯乙烯、噁嗪或酞菁。它们总是含有增溶基团（主要是磺酸基，少部分为羧基和羟基）。直接染料用于棉、人造丝、亚麻、黄麻、丝绸和聚酰胺纤维的染色。以下化学物质和助剂常用于直接染料的染色：

- 电解质：通常是氯化钠或硫酸钠。抑制纤维表面的负电位，并有助于染料耗尽，也有利于纤维上染料离子的聚集。
- 后处理剂：用于提高湿牢度。通常是具有长烃链的季铵化合物。甲醛与胺、单氰胺或双氰胺以及多核芳香酚的缩合产物也可用作后处理剂。

直接染料的研究重点是取代致癌的联苯胺染料。

(9) 金属配合物染料

金属配合物染料可大致分为两类：1∶1金属配合物和1∶2金属配合物。染料分子通常是含有附加基团如羟基、羧基或氨基的单偶氮结构，可与过渡金属离子形成强配位配合物。常用金属离子有铬、钴、镍和铜。用于金属配合物染料的三价铬和其他过渡金属是生色团的组成部分。金属配合物染料对蛋白质纤维有良好的附着力。1∶2金属配合物染料也适用于聚酰胺纤维。使用金属配合物染料染色需要使用以下化学品和助剂：

- pH调节剂：硫酸、甲酸和乙酸。
- 电解质：硫酸钠、乙酸铵和硫酸盐。
- 匀染剂：阴离子和非离子表面活性剂的混合物。

染料的生物去除率小于50%。有些染料废水中含有有机卤素。并且由于染料不固定，所以在废水中也可以找到金属元素。

1.3 三苯基甲烷染料

1.3.1 三苯基甲烷染料的发现

Perkin偶然发现一种紫红色染料是不纯苯胺重铬酸盐氧化的产物，这促使化学家们用一系列试剂来检测苯胺的氧化。1858—1859年间，法国化学家弗朗索瓦·埃马纽埃尔·维金（François-Emmanuel Verguin）发现苯胺与四氯化锡反应产生了一种紫红色或玫瑰色染料，他将其命名为品红，它是第一种三苯基甲烷染料。在品红制剂中无意中加入过量的苯胺导致了苯胺蓝的发现，这是一种有前途的新染料，尽管它的水溶性很差。根据这些染料的分子式，Hoffmann认为苯胺蓝是带有三个以上苯基的品红（$—C_6H_5$），但化学结构仍然未知。在一项细致的研究中，英国化学家爱德华·钱伯斯·尼科尔森（Edward Chambers Nicholson）证明纯苯胺不产生染料。Hoffmann表明，必须存在甲苯胺才能产生这些染料。所有这些染料，包括品红，都是由含有未知量甲苯胺的苯胺制备的。

1878 年，德国化学家埃米尔·费歇尔（Emil Fischer）建立了三苯基甲烷结构，他发现对甲苯胺的甲基碳成为连接三个芳基的中心碳。品红被发现是对硝基苯胺的混合物，即碱性红 9 和一个氨基（—NH_2）邻位有一个甲基（—CH_3）的同系物；它的经典名称是碱性紫 14。苯胺蓝中的每个氮带有一个苯基，而结晶紫中的每个氮都被二甲基化。孔雀石绿与结晶紫的区别在于具有一个未取代的芳基。这些早期合成的染料具有几个不同的名称也不足为奇。例如，孔雀石绿也被称为苯胺绿、中国绿和苯甲醛绿，是基本绿 4（C. I. 42000），并有十几个其他商品名。

Nicholson 独立发现了苯胺蓝，并发现用硫酸处理可以大大增大其水溶性。在芳基环上加入磺酸基团（—SO_3H）的过程被发现适用于许多染料，并成为提高水溶性的标准方法。颜色指数中列出的几百种三苯基甲烷染料中的大多数是在 1900 年以前合成的。在一些情况下，一个苯环被萘基取代，其取代基包括—NH_2、—OH、—SO_3Na、—COOH、—NO_2、—Cl 和烷基。虽然大多数取代基充当增色剂的角色，但磺酸盐的存在只是为了提高染料的溶解度，这也可通过氨基及其盐酸盐和羟基得到改善。

1.3.2 三苯基甲烷染料的结构

19 世纪 50 年代后期合成了第一批三苯基甲烷染料。品红就是一个例子，它由氯乙烯与苯胺反应制得。三苯基甲烷染料有几种不同的结构，但没有一种能精确解释观察到的光谱特征[22,23]。因此，三苯基甲烷染料通常被认为是共振杂合体。然而，为了方便起见，通常只显示一个杂化，如结晶紫（碱性紫 3），其最大波长为 589nm。中心碳原子周围的邻氢原子显示出相当大的空间重叠。因此，可以假定染料中的三个苯基不是共面的，而是以一种类似于三叶螺旋桨的方式扭曲[24]。三个苯基的对位取代决定了染料的色调。当仅存在一个氨基时，如在盐酸品红胺中，最大光吸收在 440nm，色调为淡橙黄色。然而，当在不同的环中存在至少两个或更多个氨基时，共振的可能性大大增加，导致更大的吸收强度和向更长波长发生强烈的红移。例如多伯纳紫（Doebner's violet）在 562nm，呈现红紫色；而对硝基苯胺在 538nm，呈现是蓝紫色。具有商业价值的氨基衍生物包含两个或三个氨基。随着 *N*-烷基化增强了伯胺的碱性，观察到进一步的强烈红移。例如，孔雀石绿（碱性绿 4），吸收峰在 621nm。随着苯基化程度的增大，伯氨基的苯基化也导致吸收波长的红移增加。

取代基对三苯基甲烷染料颜色和组成的空间效应已被广泛研究[25]。用甲基取代结晶紫（最大光吸收 589nm）中与中心碳原子邻接的氢原子，会导致向三甲基衍生物的均匀红移，并降低吸收率值[26]。这些现象表明，调节邻甲基所需的轴向旋转调节被三个苯环统一分担。然而，2,6-二甲基衍生物（最大光吸收 635nm）显示出每个甲基更大的红移。二甲氨基苯环经历了大部分的旋转扭曲，通过绕中心键扭曲来减轻空间应变，使得电荷位于另外两个二甲基氨基苯基环上[23]。

1.3.3 三苯基甲烷染料的化学性质

一般染料，特别是三苯基甲烷染料，一旦生成就很少再进行化学加工。取代基的引入通常在中间体的制造过程中进行，其中引入基团的位置和数量可以更精确地控制。染料在使用过程中和使用之后有时会暴露在氧化和还原条件下。

(1) 氧化

虽然许多三苯基甲烷染料是通过氧化无色碱制备的，但也容易被强氧化剂破坏。需要仔细选择氧化剂和反应条件，以防止在制造阶段产品的损失。孔雀石绿过度氧化会产生醌亚胺[27]。过度氧化也可能导致烷基从氨基取代基上氧化裂解。因此，三苯基甲烷染料容易被次氯酸钠破坏，限制了它们作为纺织染料的用途。

三苯基甲烷染料对光化学氧化极为敏感，这是它们在天然纤维上耐光性差的原因[28]。影响三苯基甲烷染料在天然和合成纤维上褪色（降解）速度的因素很多，包括基底纤维的类型、染料所附着的纤维结合位点（如磺酸或羧酸基团）的性质、氧的作用，以及水的硬度及酸碱度（程度较小）。纤维素基质上孔雀石绿的光降解产物被鉴定为二苯甲酮和 4-二甲氨基苯甲酮[29]。有人提出，三苯基甲烷染料的分解发生在染料结合处，产生的甲醇化染料对紫外线辐射有吸收。激发态甲醇要么发生自由基裂解，然后与水和氧反应，要么直接与水和氧反应生成上述产物。

数项研究表明，*N*-脱烷基化与裂解同时发生，并有助于光降解。由于分子中存在 *N*-烷基，所以将取代基引入孔雀石绿的苯环中并没有显著改善耐光性。在吲哚基二苯基甲烷染料的类似物中，用羊毛固蓝 FBL 中的 *N*-芳基取代，在评价颜色变化的 1～8 灰度级上可将耐光性等级提高 1～2 个点。在空气和水分存在下辐射 *N*-烷基可将其转化为醛，如甲醛或乙醛。这种 *N*-脱烷基反应是染料光化学中的一种普遍现象，在噻嗪染料[30]、罗丹明染料[31]和 *N*-甲基氨基蒽醌[32]中都有报道。

(2) 还原

三苯基甲烷染料可通过多种试剂轻松还原为无色形式，包括亚硫酸氢钠、酸(盐酸、乙酸)、锌粉和氨以及浓盐酸中的氯化钛。用三氯化钛还原（Knecht 方法）可快速测定三苯基甲烷染料［式(1-2)］。进行滴定至无色终点，通常非常敏锐。

$$Ar_3COH+2TiCl_3+2HCl = Ar_3CH+2TiCl_4+H_2O \quad (1\text{-}2)$$

(3) 磺化

烷基氨基三苯基甲烷染料的直接磺化反应得到取代产物的混合物。尽管含有苯胺基或苄基氨基的染料具有更高的取代选择性，但磺化中间体如 3[(*N*-乙基-*N*-苯基氨基)甲基]苯磺酸（乙基苄基苯胺磺酸）是优选的起始原料。磺化程度取决于反应条件。单磺化衍生物，通常被称为碱性蓝，如 C.I. 酸性蓝 119，以钡盐或钙盐形式用在印刷品中。二磺化化合物，如 C.I. 酸性蓝 48，使用其钠盐或铵盐来使纸张变蓝。而三磺酸衍生物或油墨蓝，如 C.I. 酸性蓝 93，用于书写油墨。

(4) *N*-烷基化和 *N*-芳基化

含有高度烷基化氨基的染料是一般由高度烷基化的中间体制备，而不是通过带有伯氨基染料的直接烷基化制备。然而，4,4′,4″-三氨基三苯基甲烷（对氨基苯甲酸）可以与过量的苯胺和苯甲酸进行 *N*-苯基化，得到绿蓝色的 *N*,*N*′,*N*″-三苯基氨基三苯基甲烷盐酸盐和 C.I. 溶剂蓝 23（$\lambda_{max}=586nm$）。较短的反应时间和使用较少量的苯甲酸会产生更多的红蓝色，生成单芳基化和二芳基化产物的混合物。

(5) 生成颜料

三苯基甲烷染料可以转化成两种不溶性化合物，在工业上用作颜料[33]，两者都是三苯基甲烷染料的盐。水溶性阳离子染料与磷钼酸、磷钨酸、铁氰化铜结合，偶尔与硅钼酸和磷钨酸结合形成不溶性配合物。这些配合物被称为颜料色淀，可提供干净、鲜艳的红色和紫色调。这些颜料用于印刷油墨，特别是包装和特种印刷油墨。衍生自三苯基甲烷染料的第二种颜料被称为碱性蓝。商业上重要的颜料来源于二丙烯酸酯或三丙烯酸酯玫瑰苯胺，如碱性蓝 61，可以通过二苯基甲烷碱法或苯并三氯法制备。制造水不溶性单磺酸盐需要浓硫酸。碱性蓝的主要用途是用作基于炭黑的油墨中的着色颜料，其中需要廉价的蓝色成分来校正基础颜料的天然棕色调。应用的主要领域是印刷油墨，尤其是胶印和凸版印刷。用这种颜料给打印机色带上色至今仍在使用。

1.3.4 三苯基甲烷染料的应用

三苯基甲烷染料的应用主要局限于非纺织品，大量用于制备印刷油墨、浆料和纸张印刷行业的有机颜料。在这些行业中，着色的成本和亮度比耐光性更重要。三苯基甲烷染料及其无色前体，如甲醇和内酯，广泛应用于高速光耦合和光成像系统的热敏、光敏和压敏记录材料，以及印刷板和集成电路的生产。三苯基甲烷染料也有一些特殊应用，例如汽车防冻液、卫生间制品着色、复写纸生产、打字机色带的油墨以及高速打印机的喷墨打印。

除天然纤维和腈纶的染色和印刷外，三苯基甲烷染料还适用于其他基材的着色，例如纸张、陶瓷、皮革、毛皮、阳极氧化铝、蜡、抛光剂、肥皂、塑料、药物和化妆品。有几种三苯基甲烷染料被用作食品着色剂，并在严格的加工控制下生产，一般情况下是灿烂绿和蓝色染料，但红色和紫色染料也可用于食品着色。三苯甲烷染料也广泛用作微生物染色剂[34]。一些三苯基甲烷衍生物是非常有效的羊毛防蛀剂，作为感光材料的防晕染料和指示剂在文献中也有提及[35]。

三苯基甲烷染料可用于玻璃着色。使用水溶性染料，例如酸性蓝 83，可以通过光刻技术制备彩色滤光片。可以在玻璃上建立三原色的红色、绿色和蓝色矩阵，以生产滤光片。彩色滤光片用于生产平板电视。三苯基甲烷染料的其他高科技应用包括电子照相和光学数据存储。随着使用硒的影印机数量的减少，有机光电导体，尤其是正电荷控制剂变得越来越重要。三苯基甲烷染料，如双(三甲基硅烷基)氨基钾［potassium bis(trimethylsilyl)amide］便是其中一种。三苯基甲烷染料的吸收光谱可以扩展到近红外区域，因此可作为光信息记录介质的红外吸收剂。

三苯基甲烷染料对革兰氏阳性细菌和酵母菌具有较强的抗菌作用。此外，在皮肤和黏膜上，这些染料具有抗增殖和表面置换作用，在治疗湿疹方面具有理想的干燥效果[36]。因此，三苯基甲烷染料水溶液比酒精溶液更早应用于成人，而且还被广泛应用于儿童风疹和鹅口疮的治疗[36]。鉴于在经过治疗的真菌性改变区域上形成强烈的色彩对比，目前还要求在耳道中使用灿烂绿[37]。世界卫生组织将廉价的 0.5%龙胆紫溶液分类为皮肤病不可或缺的药物[38]。在美国和加拿大，虽然官方未对龙胆紫进行过儿童特别检查，但提到了其在儿童口腔中的使用以及哺乳期间的乳头护理[38]。除了使用溶液和乳霜制剂治疗皮肤病之外，还建议将其用于念珠菌性阴道炎的治疗[39]。曙红钠盐在 pH 中性时几乎没有抗菌性能，仅在 pH 值约为 5 时以水溶性较差的酸性形式达到饱和极限可用于皮肤病的治疗。酸性曙红溶液具有干燥特性，可用于轻微擦伤，在预防和治疗皮肤感染中具有辅助作用[40]。

1.4 三苯基甲烷染料的危害

染料主要通过其生产和使用过程中产生的废水进入环境。尽管已经报道了从生产某些三苯基甲烷染料的过程中释放到环境中的物质清单，但尚无法获得释放到环境中的染料的准确数据[41]。然而，各种估计和计算表明，染料的1%～2%在制造阶段损失，1%～10%在使用阶段损失[42]。据估计，废气和洗涤液中会损失2%～3%的碱性染料。

孔雀石绿在鱼类中的存在被确定为一种健康风险[43]。目前，孔雀石绿仍然在一些没有严格法律的国家使用。2002年，我国农业部将其列入《食品动物禁用的兽药及其它化合物清单》，禁止在食品动物中使用。虽然如此，但目前关于销售和购买并无限制。三苯基甲烷染料中对环境危害最大的是结晶紫和孔雀石绿。

1.4.1 结晶紫

结晶紫（N,N,N',N',N'',N''-六甲基对玫瑰苯胺），是一种三苯基甲烷染料，作为一种生物着色剂和纺织加工业中的纺织染料，已被广泛应用于人类和兽医医学[44,45]。结晶紫也称为龙胆紫（不纯的形式），是一种阳离子染料，每个苯环上有一个二甲氨基（图1-3）。结晶紫被广泛用作棉花和丝绸等纺织品的紫色

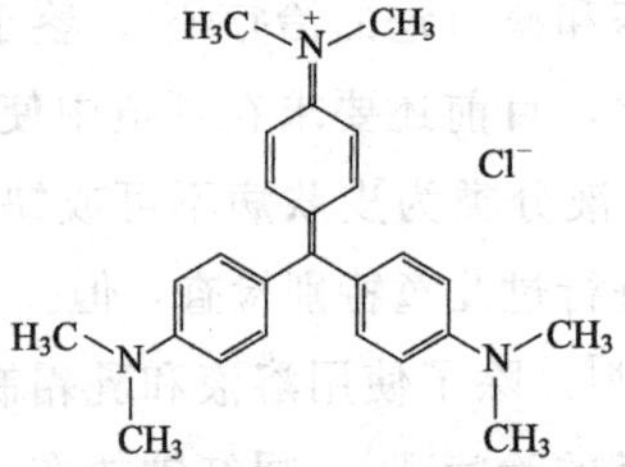

图1-3 结晶紫的化学结构式

染料，并为涂料和油墨提供深紫色。结晶紫还用于尼龙、改性尼龙和羊毛的染色，以及塑料、汽油、清漆、脂肪、油和蜡的着色[46-48]。结晶紫还可用作医用溶液中的诱变剂和抑菌剂，以及用作预防禽类饲料中真菌生长的抗微生物制剂[49,50]。此外，医学界还将结晶紫用作生物染色剂，并且是革兰氏染色剂的重要成分。这种染料还可用作人体皮肤外用消毒剂。由于蛋白质是由不同的氨基酸组成的，很容易被结晶紫染色，因此可用作血液“指纹”的增强剂。

据报道，结晶紫是一种难降解的有机染料，长期存在于环境中，对水生和陆地生物都有毒性作用[44,45]。体外研究表明，结晶紫在某些鱼类中起到有丝分裂毒药、强致癌物、强破纤维原和促进肿瘤生长的作用[44,51,52]。因此，结晶紫被认为是一种生物危害物质。由于结晶紫是一种阳离子染料，对哺乳动物细胞具有剧毒性，因此可引起中度眼部刺激，对光致敏疼痛，对角膜和结膜造成永久性损伤。并且，在极端情况下，它也可能导致呼吸衰竭和肾衰竭[53,54]。三苯基甲烷染料是使用最广泛的皮肤病药物之一。早先，结晶紫被广泛用于通过口服途径治疗蛲虫病，并在人类和家畜中局部应用。已证明结晶紫在不同条件下能有效地控制真菌生长，因此添加到家禽饲料中。通过广泛的药用和商业用途，使人们直接或间接地将自身暴露于结晶紫中[44,54]。

来自不同行业的含有结晶紫的深色废水，由于阳光穿透力降低，对水生植物的光合作用有显著影响。并且，由于芳香族、金属和氯化物等的存在，也可能对其他水生生物有毒[55,56]。在受纳水体表面形成的一层较薄的深色染料最终会降低水体中的溶解氧含量和水生植物的光合作用。而在农业土壤中，结晶紫会抑制作物的种子萌发和生长[57]。

由于结晶紫对人体健康的不利影响，其已被列为危险化学品，并已被禁止用于水产养殖和食品工业。然而，由于其相对较低的成本、易得性和有效性，所以在一些领域仍然使用[58]。染料工业在染色过程中约有10%～20%的染料以游离态排放，但如果使用偶氮染料，该值可能会增加到50%。大量的染料和废水排放到水资源中可能会严重影响美观、气体溶解度和水透明度[59,60]。因此，从不同行业的废水中去除结晶紫，不仅对保护人类健康，而且对保护土壤和水生态系统都是至关重要的。从纺织废水中去除结晶紫的方法多种多样，包括化学氧化还原法、物理沉淀絮凝法、光解法、吸附法、电化学处理法、深度氧化法、反渗透法和生物降解法[45]。在这些处理方法中，生物降解法被认为具有成本效益、结构简单、易于操作、代谢途径多样和微生物多样性等诸多优点。与物理和化学处理方法相比，它具有更广泛的应用和环境友好性，并且产生的污泥量更少[61-63]。

1.4.2 孔雀石绿

孔雀石绿是一种碱性染料，易溶于水，其化学结构式如图 1-4 所示。孔雀石绿的杀菌作用自 20 世纪 30 年代中期就已为人所知[64]。在 20 世纪 50 年代，孔雀石绿作为一种非常有效的防腐剂，被用来对抗体内和体外的寄生虫。在 20 世纪 60 年代，孔雀石绿被证明是对抗原生动物体外寄生虫，特别是多裂小瓜虫最有效的方法。当孔雀石绿对鱼卵中的水真菌腐霉的有效性[65,66]及其在鲑鱼增殖性肾病治疗中的适用性[67]得到证明时，孔雀石绿变得更加重要。孔雀石绿最常被用于治疗龙线虫病、滴虫病和皮肤真菌感染。短期孔雀石绿浴也被推荐作为鲑科鳃黄杆菌病的治疗方法[68]。

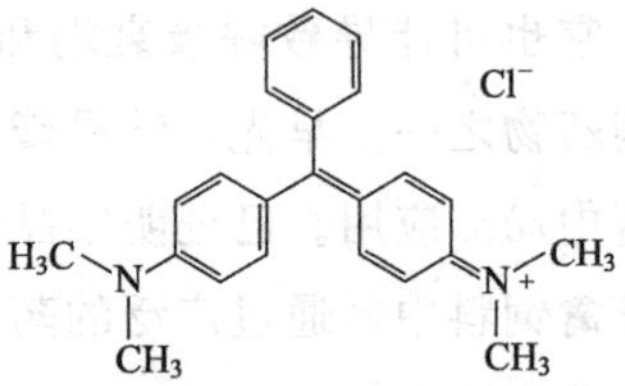

图 1-4 孔雀石绿的化学结构式

孔雀石绿残留物在环境中持续存在，并对一系列水生和陆地动物造成急性毒性，对公众健康造成严重风险，并造成潜在的环境问题。据报道，孔雀石绿对鱼类有致癌、致畸、繁殖力降低和呼吸道毒性等效应。血液的某些生化参数也会发生显著变化，如总胆固醇增加，血浆中磷和钙水平降低等[69]。哺乳动物的临床和实验研究表明，孔雀石绿可作为一种多器官毒性化合物。例如：兔子肾脏改变，大鼠生长和繁殖力降低；肝、脾、肾和心脏损害；皮肤、眼睛、肺和骨骼病变；对大鼠和小鼠的致畸作用。在涉及肝、肺、乳腺、卵巢和甲状腺肿瘤的实验动物中证实了致癌作用[70,71]。除了所呈现的毒性效应之外，从患病的鲤鱼和鳟鱼中分离出的一些细菌对孔雀石绿表现出抗性，这是孔雀石绿造成环境问题的证据[70]。尤其是考虑到这种染料对食品制备中常用的热处理稳定时，食品中孔雀石绿的残留变得更加令人担忧。Mitrowska 等研究了使用不同烹饪方法的鲤鱼片中孔雀石绿和隐性孔雀石绿的持久性[72]。烹饪 15min 后，观察到隐性孔雀石绿的浓度并未降低，这表明该物质在食品的热处理条件下具有更高的稳定性。2007 年，孔雀石绿被联合国粮食及农业组织（FAO）/世界卫生组织（WHO）食品添加剂联合专家委员会（JECFA）列入毒理学评估优先事项清单，目的是为食品中兽药残留分法典委员会（CCRVDF）提供关于在供人食用的动物中使用该物质的指导［美国食品药品监督管理局（USFDA），2007 年］。在 CCRVDF 第 18 届会议的报告中，在审查了现有文献后，JECFA 重申，由于其主要生物转化产品隐性孔雀石绿具有毒性，孔雀石绿不能用于人类食用的动物。

三苯基甲烷染料由于其颜色鲜艳、性质稳定，故而在多种行业广泛应用，也造成了显著的环境问题。因此，如何处理三苯基甲烷染料污染，已成为巨大的现实挑战和亟须解决的问题。

参 考 文 献

［1］ Clark R J H，Cooksey C J，Daniels M A M，Withnall R. Indigo，woad，and Tyrian Purple：important vat dyes from antiquity to the present. Endeavour，1993，17（4）：191-199.

［2］ Anon. William Henry Perkin. Wikipedia，2020.

［3］ Ball P. Perkin，the mauve maker. Nature，2006，440（7083）：429-429.

［4］ 太平洋证券．十五年染料周期大复盘，环保趋严下的新寡头格局——染料深度报告．2018-01-31.

［5］ 财通证券．染料行业有望迎来新一轮复苏——染料行业专题报告．2018-02-23.

［6］ de las Marías P M. Química y física de las fibras textiles. Alhambra，1976.

［7］ Van der Zee F P，Lettinga G，Field J A. Azo dye decolourisation by anaerobic granular sludge. Chemosphere，2001，44（5）：1169-1176.

［8］ Gomes J R. Estrutura e propriedades dos corantes. Barbosa e Xavier Lda，Braga，2001.

［9］ Guaratini C C I，Zanoni M V B. Textile dyes. Química Nova，2000，23（1）：71-78.

［10］ Ramalho P A. Degradation of dyes with microorganisms：studies with ascomycete yeasts. 2005.

［11］ Ingamells W. Colour for textiles：a user's handbook. Society of dyers and colourists，1993.

［12］ Forgacs E，Cserhati T，Oros G. Removal of synthetic dyes from wastewaters：a review. Environment international，2004，30（7）：953-971.

［13］ Vankar P S. Chemistry of natural dyes. Resonance，2000，5（10）：73-80.

［14］ Chengaiah B，Rao K M，Kumar K M，Alagusundaram M，Chetty C M. Medicinal importance of natural dyes—a review. International Journal of PharmTech Research，2010，2（1）：144-154.

［15］ Shahid M，Shahid-ul-Islam，Mohammad F. Recent advancements in natural dye applications：a review. Journal of Cleaner Production，2013，53：310-331.

［16］ Siva R. Status of natural dyes and dye-yielding plants in India. Current science，2007M，92（7）：916-925.

［17］ Singh R，Jain A，Panwar S，Gupta D，Khare S K. Antimicrobial activity of some natural dyes. Dyes and Pigments，2005，66（2）：99-102.

［18］ Chavan R B. Environmentally friendly dyes. Clark M. Handbook of Textile and Industrial Dyeing：Vol 1. Cambridge：Woodhead Publishing，2011：515-561.

［19］ Lucas M S，Dias A A，Sampaio A，Amaral C，Peres J A. Degradation of a textile reactive Azo dye by a combined chemical-biological process：Fenton' s reagent-yeast. Water research，2007，41（5）：

1103-1109.

[20] Zollinger H. Color chemistry: syntheses, properties, and applications of organic dyes and pigments. New York: John Wiley and Sons, 2003.

[21] Robinson T, McMullan G, Marchant R, Nigam P. Remediation of dyes in textile effluent: a critical review on current treatment technologies with a proposed alternative. Bioresource technology, 2001, 77 (3): 247-255.

[22] Bury C R. Auxochromes and resonance. Journal of the American Chemical Society, 1935, 57 (11): 2115-2117.

[23] Pauling L. A theory of the color of dyes. Proceedings of the National Academy of Sciences of the United States of America, 1939, 25 (11): 577.

[24] Beckett A H. Stereochemical factors in biological activity//Fortschritte Der Arzneimittelforschung/Progress in Drug Research/Progrès Des Recherches Pharmaceutiques. Berlin: Springer, 1959: 455-530.

[25] Kitao T. Novel syntheses and characteristics of functional heteroaromatics. Chemical Resources: New Developments in Organic Chemistry: a Report on a Special Research Project, 1988: 298.

[26] Gandhi S S, Hallas G, Thomasson J. Steric and electronic effects in basic dyes: Ⅴ-electronic absorption spectra of derivatives of malachite green containing methoxy groups in the phenyl ring. Journal of the Society of Dyers and Colourists, 1977, 93 (12): 451-454.

[27] Gray G W. Steric effects in conjugated systems. Pittsburgh: Academic Press, 1958.

[28] Kuramoto N, Kitao T. The contribution of singlet oxygen to the photofading of triphenylmethane and related dyes. Dyes and Pigments, 1982, 3 (1): 49-58.

[29] Porter J J, Spears Jr S B. The photodecomposition of Cl Basic Green 4. Textile Chemist and Colorist, 1970, 2 (11): 33.

[30] Usui Y, Obata H, Koizumi M. Photoreduction of methylene blue by visible light in the aqueous solution containing certain kinds of inorganic salts. I. General features of the reaction. Bulletin of the Chemical Society of Japan, 1961, 34 (8): 1049-1056.

[31] Evans N A. Photofading of rhodamine dyes Ⅱ-photode-alkylation of rhodamine B. Journal of the Society of Dyers and Colourists, 1973, 89: 332.

[32] Giles C H, Sinclair R S. Photodecomposition of aminoanthraquinone disperse dyes on poly (ethylene terephthalate) . Journal of the Society of Dyers and Colourists, 1972, 88 (3): 109-113.

[33] Herbst W, Hunger K. Industrial organic pigments: production, properties, applications. New York: John Wiley and Sons, 2006.

[34] Sabnis R W. Handbook of biological dyes and stains: synthesis and industrial applications. New York: John Wiley and Sons, 2010.

[35] Chalkley L. Photometric papers sensitive only to short wave ultraviolet. JOSA, 1952, 42 (6): 387-392.

[36] Nürnberg W, Reimann H. Nutzen-Risiko-Abwägung bei rezeptur der triphenylmethanfarbstoffe. Der Hautarzt, 2008, 59 (10): 833-837.

[37] Michel O. Therapieziel "Trockenlegung"—Das chronisch "laufende Ohr". HNO Nachrichten，2001，5：38-42.

[38] Maley A M，Arbiser J L. Gentian Violet：A 19th century drug re-emerges in the 21st century. Experimental Dermatology，2013，22 (12)：775-780.

[39] Micromedex I B M. Gentian violet (Vaginal Route) proper use. Mayo Clinic，2020.

[40] Fischer H，Reimann H. Eosin vielleicht überflüssig，aber nicht bedenklich. Pharm Ztg，2000，145：509-510.

[41] US EPA. Dyes and pigments production wastes—waste from the production of dyes and pigments listed as hazardous. 2005.

[42] Laing I G. The impact of effluent regulations on the dyeing industry. Review of Progress in Coloration and Related Topics，1991，21 (1)：56-71.

[43] Andersen W C，Turnipseed S B，Roybal J E. Quantitative and confirmatory analyses of malachite green and leucomalachite green residues in fish and shrimp. Journal of agricultural and food chemistry，2006，54 (13)：4517-4523.

[44] Au W，Pathak S，Collie C J，Hsu T C. Cytogenetic toxicity of gentian violet and crystal violet on mammalian cells in vitro. Mutation Research/Genetic Toxicology，1978，58 (2)：269-276.

[45] Azmi W，Sani R K，Banerjee U C. Biodegradation of triphenylmethane dyes. Enzyme and microbial technology，1998，22 (3)：185-91.

[46] Gregory P. Dyes and dye intermediates//Kirk-Othmer Encyclopedia of Chemical Technology. New York：American Cancer Society，2000.

[47] Parshetti G，Kalme S，Saratale G，Govindwar S. Biodegradation of malachite green by *Kocuria rosea* MTCC 1532. Acta Chimica Slovenica，2006，53 (4)：492-498.

[48] Daneshvar N，Ayazloo M，Khataee A R，Pourhassan M. Biological decolorization of dye solution containing Malachite Green by microalgae *Cosmarium* sp. Bioresource Technology，2007，98 (6)：1176-1182.

[49] Littlefield N A，Blackwell B—N，Hewitt C C，Gaylor D W. Chronic toxicity and carcinogenicity studies of gentian violet in mice. Fundamental and Applied Toxicology，1985，5 (5)：902-912.

[50] Mittal A，Mittal J，Malviya A，Kaur D，Gupta V K. Adsorption of hazardous dye crystal violet from wastewater by waste materials. Journal of Colloid and Interface Science，2010，343 (2)：463-473.

[51] Cho B P，Yang T，Blankenship L R，Moody J D，Churchwell M，Beland F A，Culp S J. Synthesis and characterization of *N*-demethylated metabolites of malachite green and leucomalachite green. Chemical Research in Toxicology，2003，16 (3)：285-294.

[52] Fan H J，Huang S T，Chung W H，Jan J L，Lin W Y，Chen C C. Degradation pathways of crystal violet by fenton and fenton-like systems：condition optimization and intermediate separation and identification. Journal of Hazardous Materials，2009，171 (1)：1032-1044.

[53] Ahmad R. Studies on adsorption of crystal violet dye from aqueous solution onto coniferous pinus bark powder (CPBP) . Journal of Hazardous Materials，2009，171 (1)：767-773.

[54] Amini M, Younesi H. Biosorption of Cd(Ⅱ), Ni(Ⅱ) and Pb(Ⅱ) from aqueous solution by dried biomass of *Aspergillus niger*: Application of response surface methodology to the optimization of process parameters. CLEAN-Soil, Air, Water, 2009, 37 (10): 776-786.

[55] Gill P K, Arora D S, Chander M. Biodecolourization of azo and triphenylmethane dyes by *Dichomitus squalens* and *Phlebia* spp. Journal of industrial microbiology and biotechnology, 2002, 28 (4): 201-203.

[56] Liu W, Chao Y, Yang X, Bao H, Qian S. Biodecolorization of azo, anthraquinonic and triphenylmethane dyes by white-rot fungi and a laccase-secreting engineered strain. Journal of industrial microbiology and biotechnology, 2004, 31 (3): 127-132.

[57] Kalyani D C, Patil P S, Jadhav J P, Govindwar S P. Biodegradation of reactive textile dye Red BLI by an isolated bacterium *Pseudomonas* sp. SUK1. Bioresource Technology, 2008, 99 (11): 4635-4641.

[58] Schnick R A. The impetus to register new therapeutants for aquaculture. The Progressive Fish-Culturist, 1988, 50 (4): 190-196.

[59] Moturi B, Charya M S. Decolourisation of crystal violet and malachite green by fungi. Science World Journal, 2009, 4 (4): 28-33

[60] Shah M P, Patel K A, Nair S S, AM D. Microbiological removal of crystal violet dye by *Bacillus subtilis* ETL-2211. OA Journal of Biotechnology (UK), 2013, 2 (1): 9.

[61] Banat I M, Nigam P, Singh D, Marchant R. Microbial decolorization of textile-dyecontaining effluents: A review. Bioresource Technology, 1996, 58 (3): 217-227.

[62] Méndez-Paz D, Omil F, Lema J M. Anaerobic treatment of azo dye Acid Orange 7 under fed-batch and continuous conditions. Water Research, 2005, 39 (5): 771-778.

[63] Pandey A, Singh P, Iyengar L. Bacterial decolorization and degradation of azo dyes. International Biodeterioration and Biodegradation, 2007, 59 (2): 73-84.

[64] Foster F J, Woodbury L. The use of malachite green as a fish fungicide and antiseptic. The Progressive Fish-Culturist, 1936, 3 (18): 7-9.

[65] Oláh J, Farkas J. Effect of temperature, pH, antibiotics, formalin and malachite green on the growth and survival of Saprolegnia and Achlya parastic on fish. Aquaculture, 1978, 13 (3): 273-288.

[66] Alderman D J. Malachite green: a review. Journal of Fish Diseases, 1985, 8 (3): 289-298.

[67] Clifton-Hadley R S, Alderman D J. The effects of malachite green upon proliferative kidney disease. Journal of Fish Diseases, 1987, 10 (2): 101-107.

[68] Citek J, Svobodova Z, Tesarcik J. General prevention of fish diseases. Diseases of freshwater and aquarium fish (in Czech). Prague: Informatorium, 1997: 9-49.

[69] Srivastava S J, Singh N D, Srivastava A K, Sinha R. Acute toxicity of malachite green and its effects on certain blood parameters of a catfish, Heteropneustes fossilis. Aquatic toxicology, 1995, 31 (3): 241-247.

[70] Srivastava S, Sinha R, Roy D. Toxicological effects of malachite green. Aquatic toxicology, 2004,

66 (3): 319-329.

[71] Culp S J, Mellick P W, Trotter R W, Greenlees K J, Kodell R L, Beland F A. Carcinogenicity of malachite green chloride and leucomalachite green in B6C3F1 mice and F344 rats. Food and Chemical Toxicology, 2006, 44 (8): 1204-1212.

[72] Mitrowska K, Posyniak A, Zmudzki J. The effects of cooking on residues of malachite green and leucomalachite green in carp muscles. Analytica chimica acta, 2007, 586 (1-2): 420-425.

第2章
三苯基甲烷染料的脱色

2.1 脱色方法

三苯基甲烷染料应用广泛，而且可以富集，在印染废水中普遍存在，严重污染水生环境。为了从水体中去除这类染料，已经开发了一系列物理、化学和生物方法[1-5]。虽然化学和物理法（如吸附、沉淀、混凝、过滤、电解、光降解和化学氧化）对脱色是有效的，但存在成本高、效率低、通用性有限、干扰其他废水成分、形成有害代谢物和高强度能源需求大等不足[6,7]。废水中三苯基甲烷染料的生物降解因过程简单、不产生二次污染和剩余污泥，是一种非常有效且环境友好的方法[8]。

2.1.1 物理法

(1) 吸附法

吸附是去除染料最有效、最经济可行的技术之一。尽管无机载体具有良好的力学和化学稳定性、高比表面积和抗微生物降解性，但有机载体在可再生方面具有优势，而且属于工业过程废物，几乎没有商业价值[9]。

在无机载体中，碳基无机载体以及其他材料已经被用于吸附各种类型的染料。影响染料在特定载体（如活性炭）上吸附的关键参数包括染料的分子大小和溶解度。水溶性亲水染料吸附性差的一个关键原因是染料的极性与碳的非极性相斥[10]。低分子量的酸性和活性染料的吸附性低，而分子量较高的碱性和直接染料吸附性更高。分散染料的分子量中等偏高，但具有疏水性。分散染料、还原染料和颜料的低溶解度和胶体分散特性导致其在室温下对碳的吸附速率较慢。

为了更好地理解碳表面吸附的物理化学过程，均匀表面扩散模型被成功地用于描述染料的吸附[11]。已经发现褐煤基碳的吸附特性明显依赖于制备方式。磺化煤对合成染料的良好吸附特性也已得到证实。碳基载体的吸附研究结果表明，它对相当多的合成染料有很好的效果。活性炭是吸附去除染料最常用的处理方法，对阳离子、媒染剂和酸性染料的吸附非常有效，对分散、直接、还原和活性染料的吸附程度稍小[12,13]。泥炭、木屑、硅胶、粉煤灰、玉米芯和稻壳等不同材料广泛用于生产商用活性炭和去除废水中的染料。这些材料之所以具有优势，主要是因为其广泛可用性和低成本[14,15]。活性炭吸附也有一些缺点，由于过程是非选择性的，因此，废水的其他组分可能会争夺吸附位置，从而降低染料的结合能力[16]。此外，吸附过程通过将染料浓缩在表面并保持其结构不变来去除染料。当载体再生时，浓缩的染料污泥会给随后的环境处置带来问题[9]。

(2) 离子交换法

源自甘蔗渣、废纸、聚酰胺废料和甲壳质的各种离子交换树脂不仅用作脱色剂，还用作其他有机物的吸附剂[17]。从概念上讲，阴离子染料（例如酸性染料、媒染剂、活性染料、直接染料、金属配合物）以及阳离子染料（碱性染料），如果用离子交换树脂处理，可形成絮凝物形式的大型配合物，能够通过过滤分离。例如，四氯化甘蔗渣和季铵化纤维素是一种可行的离子交换树脂，能够结合水解活性染料。实验表明，高盐浓度不会破坏这种离子交换与活性染料的结合。除了水解染料与离子交换树脂季铵化位点的离子相互作用之外，其他相互作用如氢键和范德华力也可能参与其中。染料去除的效果取决于这种相互作用的类型和数量[18]。然而，由于分散染料的疏水性，这种离子交换树脂对分散染料无效[19]。

与活性炭相比，大多数离子交换树脂的流体力学性能较差，并且这些树脂难以承受迫使大量废水通过树脂床、保持高流速所需的高压。离子交换树脂的其他缺点包括在再填充过程和产生大量污泥时产生的高成本。

(3) 混凝-絮凝

混凝法也广泛应用于印染废水的脱色。尽管该工艺仅部分去除颜色，但根据所用混凝剂的不同，它也在不同程度上降低了化学需氧量（COD）。混凝作用是由颗粒表面电动电位降低以及颗粒间的缔合形成絮凝团聚体所致[20]。许多阳离子、阴离子或非离子聚合物用作混凝剂以去除染料。在混凝过程中，至少存在四个主要困难。第一，回收在混凝过程中使用的昂贵化学物质是不可行的，这些化学物质最终会被扔掉。第二，高水溶性染料抗凝结或需要使用更多混凝剂，这两个参数都显著增加成本。第三，产生彩色固废，需要另法处理。第四，污泥的毒性和处理后废水中总溶解固体（TDS）含量增加，这也可能是最严重的缺点。尽管有这些限制，但印染废水脱色在这方面的研发投入仍然有增无减。

Vijay 等采用商业 A 级废铝板作为电极，使用电凝法从废水中去除结晶紫[21]。结果表明，电凝法是处理三苯基甲烷染料废水的有效方法。实验优化了 pH、电流密度、时间、初始浓度等操作参数。在 pH 值为 11 时，在 1h 内达到平衡。经过适当的优化，在最佳电流密度为 20mA/m^{2}时，去除率达到 90%以上。适量盐的存在、恰当的反应时间和电流密度的增加均有助于提高去除率。此外，还研究了结晶紫在电凝聚过程中的吸附行为，提出了结晶紫的等温动力学模型为 Dubinin-Radushkevich 模型和拟一级模型。该过程涉及的机理被认为是化学吸附。吸附热力学研究清楚地表明，该过程是自发的、吸热的，热力学上也是有利的。

此外，用铝或亚铁/铁盐进行混凝处理后产生的污泥可能富含铝或亚铁/铁盐，如果不在处置前进行处理，可能对水生生物有毒。铁盐被认为具有较低的急性毒性，因为它们会迅速氧化成不溶性形式。然而，有报告表明，这种盐对水蚤以及鱼类的幼虫和卵都是有毒的[22-24]。还有报告称，来自饮用水处理厂的三氯化铁污泥含有高浓度的重金属，当污泥未经事先处理就排放时，会对水生生物群造成进一步的污染和长期的有害影响[25]。

(4) 膜技术

各种膜技术，包括微滤、超滤、纳滤和反渗透，已经用于在纺织工业中从流出物中回收浆料。其中某些方法用于从废水中去除颜色，而与所使用的染料类型无关。微滤膜孔径大，对废水处理没有效果。膜可以充当废水中特定成分的物理屏障，而不会降解这些成分。膜过滤有利于染料、化学品和水中杂质的去除或再利用。使用膜过滤从纺织品废水中分离出水溶性染料在文献中有详细记载。处理染料废物的膜系统的规格由许多因素决定，包括染色工艺、要去除的染料类型、废液的化学成分以及通量等[10]。用于脱色的膜必须具有合理的化学和热稳定性，允许在高通量下操作，并且能够耐受宽范围的酸碱度、温度和不同类型溶剂。

膜过滤已成为从废水中去除结晶紫的常规处理工艺的可行替代方案，并已被证明可节省运行成本和水消耗[26]。膜技术为分离染料和助染剂提供了可能性，可同时降低废水的色度和生物需氧量（BOD)/COD。通常用于处理活性染料废水，因为可减少废物量并同时回收盐[27]。该方法能够澄清、浓缩废水，最重要的是能够从流出物中连续分离结晶紫[19,28]。膜过滤技术的优点是：快速，对空间要求低，饱和液可以重复使用，但投资成本高，容易堵塞，膜的更换影响了该方法的适用性。这种技术通常在生物处理后作为第三级或最终处理工艺[16]。

(5) 辐射

伽马辐射已被用于印染废水的脱色。大多数耐化学氧化或还原过程的染料都可以通过这一过程降解。降解速率由辐射剂量和氧气供应控制[20,29]。有机物质需要足够的溶解氧才能被辐射有效分解。辐射能有效地去除活性染料、酸性染料和分散染料以及一些有毒有机化合物。

有报道研究了辐射-氯化组合工艺对纺织染料溶液的影响[30]。研究发现，伽马辐射和氯的联合处理比两种组分单独使用时的脱色效果更好。测试了几种化学染料，包括蒽醌、偶氮、金属化偶氮、硫、苯乙烯和三苯基甲烷染料。在浓度为0.2g/L时，通过加入60mg/L的辐射剂加上75mg/L的氯进行处理，可以显著提高染料溶液的最大吸收波长处的透射率。

Behnajady等使用辐射/超声波技术从水溶液中去除孔雀石绿。染料降解遵循一级动力学，羟基自由基在孔雀石绿降解中起重要作用。染料降解随着温度的升高而加快[31]。Han等使用天然沸石去除孔雀石绿。微波辐射通过吸附剂的再生来增强吸附。吸附数据采用Koble-Corrigan模型进行分析，遵循准二级动力学。在pH值为2和45℃时，最大吸附量为27.34mg/g[32]。

2.1.2 化学法

(1) 光化学法

光催化或光化学降解过程在废水处理领域越来越重要，因为染料分子可在温和条件（温度和压力）下完全矿化。光活化化学反应的特点是，在有/没有催化剂的情况下，适当能级的光子与溶液/废水中存在的化学分子相互作用引发自由基机理[33]。染料的常规化学氧化通常包括使用氧化剂，如氯、二氧化氯、臭氧、过氧化氢（H_2O_2）和高锰酸盐。该方法通过在H_2O_2存在下进行紫外线处理，使结晶紫分子降解为CO_2和H_2O[34,35]。结晶紫的降解是由高浓度羟基自由基的产生引起的，羟基自由基攻击不饱和染料分子，导致生色团的破坏，恶臭味大大减弱。然而，染料的去除率在很大程度上受到紫外线辐射强度、酸碱度、染料结构以及染液成分的影响[9,36]。

(2) 次氯酸钠法

使用次氯酸钠形式的氯的化学氧化长期以来被用于含有水溶性染料的废水处理，例如酸性染料、直接染料、金属配合物染料和活性染料纺织废水的脱色。这种方法通过氯离子攻击染料分子的氨基，并加速随后的偶氮键断裂。脱色在很大程度上受酸碱度和NaOCl浓度的影响[36,37]。然而，不溶于水的分散染料和还原染料对氯的脱色具有抵抗力[38]。另一方面，活性染料的脱色需要较长的处理时间[39]。此外，含有氨基或取代氨基的染料生色团易受快速氯脱色的影响。脱色率随着氯离子浓度的增大而增大。然而，使用高浓度的氯离子去除染料是不可取的，因为氯处理会产生有毒的氯化有机物，如对人类和环境有害的卤代烃，并增加处理水的可吸附有机卤化物（AOX）含量。一般来说，含有高分子量重氮化合物的三嗪比含有单偶氮或蒽醌的染料需要更长的时间才能完全脱色[36,40]。

(3) 臭氧化

臭氧是另一种广泛使用的氧化剂，由于与多种染料具有很高的反应活性，通常可提供良好的脱色效率[41]。它是一种极强的氧化剂，可与多种染料快速反应。臭氧分子具有选择性，会攻击发色团的不饱和键。施加到印染废水中的臭氧剂量在很大程度上取决于颜色强度和要去除的总残留 COD。臭氧的分解需要较高的 pH 值（pH＞10）。在碱性溶液中，臭氧几乎不加区别地与介质中存在的所有化合物发生反应，将它们转化为可生物降解的小分子[42-44]。因此，通过臭氧作用的脱色速度很快，但是生色团的矿化度很低，易生成有机酸、醛和酮等氧化副产物。臭氧化是一种很有前途的染料去除方法，其优点是在处理过程中不会形成化学污泥，并且除了染料之外，有机废物也可以同步去除。由于不需要其他化学物质，以及残留的臭氧很容易分解成氧气，臭氧脱色是一种效率高且无害的印染废水处理方法。

(4) 芬顿氧化

基于芬顿（Fenton）试剂的氧化体系已被广泛用于处理有机和无机污染物[45]。Fenton 试剂可用于从精炼废水中有效去除染料和有机卤化物[46]。此外，金属配合物染料中的重金属也可以在氧化铁的中和步骤中沉淀出来。Fenton 氧化过程适用于多种染料脱色，因为与臭氧化相比，它相对便宜并且可显著降低 COD[16,42]。与使用 H_2O_2 的其他方法（如絮凝、沉淀、气浮、过滤等）相比，用 Fenton 氧化法更具优势[47]。

据报道，在弱酸性（pH＝2～3）水溶液中，Fenton 产生的羟基自由基几乎完全去除了孔雀石绿的颜色[48]。例如在 pH 值为 2.6±0.2 的水溶液中，含有 1.08×10^{-5} mol/dm^3 的孔雀石绿。在温度 282～299K 下反应 1～3h，3.6×10^{-5} mol/dm^3 的 Fe^{2+} 和 8.82×10^{-5} mol/dm^3 的 H_2O_2 可让孔雀石绿的脱色率达到 98%。溶解的电解质（如 NaCl、NaBr 等）会大大降低反应速度，这表明卤化物会使羟基自由基失活。在 H_2O_2 比孔雀石绿含量高的情况下，反应遵循良好的一级动力学。但过高的 H_2O_2 让染料的整体降解速度变慢，这与一级动力学的偏离有关。

Fenton 体系中 [Fe(Ⅲ)-salen]Cl 配合物对孔雀石绿具有明显脱色作用[49]。在没有额外设备或外部能量的情况下，仅微量配合物就可让染料脱色显著增强。研究了反应参数对脱色率的影响，结果表明，[Fe(Ⅲ)-salen]Cl 配合物的增加、salen 配体与 Fe(Ⅲ) 的摩尔比的增加和 H_2O_2的初始浓度的增加都有利于脱色率的提高。染料初始浓度的增加导致脱色率下降，最佳 pH 值为孔雀绿溶液的初始 pH 值。在最佳条件下，处理 24min 后脱色率为 97.94%，总有机碳（TOC）去除率为 54.35%。在 [Fe(Ⅲ)-salen]Cl 配合物催化的 Fenton 体系中，羟基自由基是主要的活性氧化剂。孔雀绿的脱色符合伪一级动力学，活化能仅为 Fe(Ⅲ)-Fenton 体系的 49.13%。

2.1.3 生物法

总体上，各种物理化学处理方法被认为是有效的，但由于化学药品的过量使用，污泥的产生，后续处置问题，高昂的安装以及运营成本，其应用受到了限制[16,50]。因此，作为一种可行的替代方法，生物脱色因成本低廉，产生的污泥量少，对环境友好而受到越来越多的关注[51]。

使用生物质材料去除废水中的染料是一种有吸引力的选择，因为这可能会降低处理过程的总成本[52]。除生物物理吸附外，利用细菌、放线菌、酵母菌和放线菌等微生物或者酶来进行微生物脱色，有可能破坏染料的分子结构，甚至达到彻底矿化的目的[53]。因此，三苯基甲烷染料的生物脱色可通过两种方式进行：生物物理吸附/积累（染料结构保持完整）和生物转化/降解过程[54]。生物物理吸附过程可以发生在活的或死的生物上。然而，在积累过程中，染料沉积在活跃生长的细胞内[55]。这些过程的效率取决于特定染料的化学结构和浓度、吸附剂的化学性质和用量以及一些物理化学参数，如 pH 值、温度和搅拌。当含染料的废水毒性很大，而且环境不利于生物的生长和维持时，生物吸附特别有效。吸附过程很快，各种废料包括微生物废弃物，都可以用作染料吸附剂[56]。开发吸附剂再生方法和染料回收方法，使这些过程更加经济[57,58]。然而，生物物理吸附并不能消除染料的污染问题，因为去除的化合物不会被破坏，而只是被吸附截留。因此，生物转化/降解似乎更适合于染料去除[59]。生物降解不仅可以脱色，而且会产生无毒或减毒的代谢中间体。有时，可以实现染料的完全矿化（转化为二氧化碳、水和/或任何其他无机最终产物）。

2.2 生物物理吸附

生物物理吸附是印染废水脱色的主要处理方法之一。为代替昂贵的活性炭，各种低成本的天然生物材料被开发为染料吸附剂。这些生物物理吸附剂根据来源可分为：植物类、动物类和微生物类。

物理吸附机理取决于染料和吸附剂的性质。三苯基甲烷染料，如结晶紫和孔雀石绿，是一种阳离子染料，苯环上含有氨基，氮原子上带正电荷。吸附剂的不均匀表面提供了大的表面积。吸附通过物理吸附以单层形式发生，然后是多层[60-62]。单层吸附很快，之后是缓慢的颗粒内扩散过程[62]。控速步骤根据不同情况有表面吸附和粒子内扩散两种情况[62-64]。物理吸附过程也可以同时由膜传质、粒内扩散和化学反应控制[65]。

2.2.1 植物类

(1) 果皮

根据联合国粮食及农业组织数据显示，全球每年产生 13 亿吨食物垃圾，而其中约有 15%是果皮。用废弃果皮制成染料吸附材料可有效节约成本，实现废物利用。Sharma 等将马铃薯皮和印度楝树皮用六氯环己烷化学修饰后，用于去除孔雀石绿染料[56]。这两种生物吸附剂的吸附数据分别遵循弗罗因德利希(Freundlich）和焦姆金（Temkin）模型。酸处理后，吸附剂表现出化学吸附过程。数据遵循准二级动力学。马铃薯皮在 pH 值为 12 时吸附量最大；而印楝树皮在 pH 值为 2 时吸附量最大。Mary 等使用柑橘果皮去除孔雀石绿染料[66]。在 pH 值为 11、0.6g/L 剂量下，20min 足以去除 94%的染料。在 Velmurugan 等的研究中，橘皮、印度楝树叶和香蕉皮在 105℃下干燥 48h 并磨成粉末（粒径为 600μm)[67]。研究表明，橘皮对不同染料的吸附量大小依次为甲基橙＞亚甲基蓝＞罗丹明 B＞刚果红＞亚甲基紫＞氨基黑 10B。Mafra 等还制备了一种基于橘皮的吸附剂，并将其用于从合成染料废水中去除亮蓝染料[68]。他们发现当染料浓度为 30～250mg/L 时，可在 15h 的接触时间内达到平衡。橘皮吸附剂的吸附能力随着温度的升高而降低。橘皮吸附剂在 20℃、30℃、40℃、50℃和 60℃下的吸附量分别为 11.62mg/g、10.70mg/g、8.61mg/g、6.39mg/g 和 5.54mg/g。

(2) 植物叶片

Singh 等使用茶叶作为吸附剂，以去除孔雀石绿和亚甲基蓝染料[69]。将干燥的叶片粉末和浓度为 500g/L 的 H_3PO_4 溶液以 2∶1（体积比）的比例混合浸渍，并在 300℃下碳化。将碳洗涤并在 100℃下干燥，筛分至 170～200 目。在 25℃下，孔雀石绿和亚甲基蓝的单层吸附量分别为 444.44mg/g 和 454.5mg/g，吸附数据很好地拟合了朗缪尔（Langmuir）等温线。Nanda 等对藤蔓叶片进行处理，用作去除结晶紫的吸附剂[70]。将藤蔓叶粉碎并过 80 目筛。将 5 份叶粉与 3 份浓硫酸在 120～130℃下处理 24h 制备生物碳（TLR）。取部分 TLR 浸泡在 1%的碳酸氢钠溶液中，然后干燥并通过 80 目筛进行筛分，用作吸附剂（TLR-CM）。再取部分 TLR 用 5 份 1mol/L 硫酸处理 24h，然后将材料洗涤、干燥、粉化并用作吸附剂（TLR-1M）。结果表明，TLR-1M、TLR 和 TLR-CM 的结晶紫吸附量分别为 67.57mg/g、42.92mg/g 和 22.47mg/g。Chen 等用竹叶吸附孔雀绿染料[71]。该吸附剂表面含有羧基和羟基，染料去除率为 97.6%。在 pH 值为 6 条件下，60min 内染料的最大吸附量为 48.3mg/g，符合准二级动力学及 Langmuir 模型。Kushwaha 等用胡萝卜茎粉和胡萝卜叶粉去除孔雀绿[72]。吸附过程为瞬时吸热过程，控速步骤为粒内扩散和表面吸附。胡萝卜茎和叶的最大吸附量分别为 43.4mg/g 和 52.6mg/g。吸附数据符合 Freundlich 模型。Mullick 等将废弃甜叶菊叶制成活性炭，利用氢氧化钠浸渍进行改性制成吸附剂[73]。对于质量比为 1∶1 和 1∶2 的氢氧化钠改性吸附剂，染料吸附量分别为 284.45mg/g 和 288.67mg/g。这证明氢氧化钠用量越多，吸附能力越弱。数据遵循 Freundlich 模型，并很好地符合伪二级动力学。吸附过程本质上是非瞬时和吸热的。

菠萝叶粉可作为孔雀石绿脱色的潜在吸附剂。Chowdhury 等使用菠萝叶粉去除孔雀石绿[74]，发现温度在此过程中起着非常重要的作用。该生物吸附过程的活化能为 45.79kJ/mol。Arunadevi 等同样使用菠萝叶去除孔雀石绿[75]。吸附剂在马弗炉中于 1100℃活化，并被精细粉碎。最大染料去除率在 50min 内达到 90%。最适 pH 值在 2～6.5 之间。Das 等设计了一个固定床连续反应器，使用菠萝叶粉作为吸附剂去除孔雀石绿[76]。低流速条件下，物料停留时间长，因此吸附效率更高。

(3) 植物纤维

甘蔗是我国最大的制糖原料，南方蔗区甘蔗总产量 7000 多万吨，每年产生的甘蔗渣产量约为2000多万吨，具有很大的利用空间和潜力。Ho等研究了碱性紫10、碱性紫1和碱性绿4三种染料在甘蔗渣颗粒（粒径为351～589μm）上的吸附[77]。这些染料的Langmuir单层平衡容量分别为50.4mg/g、20.6mg/g和13.9mg/g。Patra等利用甘蔗渣、印度楝树叶和锯末，制备了生物吸附剂[78]。所有的生物废料都经过预处理以提高表面性能。研究发现，当使用磷酸进行处理时，去除效率为97%，远高于甲醛处理后的89%。真实废水处理实验表明，浓度为1g/100mL的吸附剂能去除79%的染料。

Hameed等使用棕榈树干纤维去除水溶液中的孔雀石绿[79]。棕榈树干自然晾干后切成约16cm³大小的碎片，研磨并过筛，获得粒径为0.5mm左右的颗粒。用沸水洗涤制备的颗粒，然后在70℃干燥24h。在pH值为10和30℃下，吸附在140min内达到平衡。染料最大去除率为82%，最大吸附量为149.35mg/g。pH值小于4时不利于染料吸附。随着酸碱度的降低，吸附量减少，原因如下：首先，酸碱度降低导致吸附剂的负电荷数量减少，带正电的表面位置数量增加。其次，当氢离子过量时，孔雀石绿的吸附能力在酸性条件下降低。Jothirani等使用超声波改性玉米芯，用于去除孔雀石[80]。在中性条件下，60min内最大单层吸附量为488.3mg/g。数据遵循准一级动力学。吸附过程是自发和放热的。颗粒内扩散不是唯一的控速步骤，边界层扩散也在一定程度上控制了吸附过程。

(4) 植物根

植物根部用于染料脱色的研究还很有限。Ren等使用大蒜根去除孔雀石绿[81]。吸附剂表面存在的羟基和羧基是染料吸附的主要原因。数据很好地符合Langmuir模型，并遵循准二级动力学。最大染料吸附量为232.56mg/g。该过程受颗粒内扩散和边界层外扩散控制。吸附过程本质上是物理过程。

(5) 花

Nethaji等用化学方法将琉璃苣花制备成活性炭，并将其用作孔雀石绿的吸附剂[82]。三种不同粒径的活性炭用于实验，分别为100μm、600μm和1000μm。最大吸附发生在pH值为6～8的范围内。在这一酸碱度范围内，吸附率变化不大，但小粒径明显吸附量更大。三种粒径的活性炭的吸附率分别为99%、95%和78%。Chukki等使用菊花去除孔雀石绿染料[83]。使用响应面方法对实验数据进行建模。在pH值为11和30℃下，75min内染料的最大去除率为99.3%。

(6) 种子

Santhi等研究了番荔枝种子对亚甲基蓝、亚甲基红和孔雀石绿染料的吸附潜力[84]。通过用H_2SO_4处理物料12h来制备活性炭。洗涤后，将物料用2%

$NaHCO_3$溶液处理以除去残留的酸，然后干燥并筛分至125～250μm。三种染料的吸附量分别为达到8.52mg/g、40.48mg/g和25.91mg/g，吸附数据与Langmuir和Freundlich等温线十分吻合。

(7) 种壳

稻壳是碾米厂的主要副产品，约为稻谷籽粒质量的20%。据了解，我国稻壳年产3800万吨，是一种良好的生物吸附剂来源。Rahman等将稻壳经磷酸、氢氧化钠和氮气处理后用于去除孔雀石绿[85]。碳化温度为500℃时，所得活性炭具有良好的去除效率，染料吸附量达到80mg/g。而吸附量也与染料浓度有关。当染料浓度从10mg/L提高到100mg/L，稻壳的吸附量也随之增加，这是因为高浓度提供了克服液固两相间的传质阻力[86]。Chowdhury等使用化学改性稻壳去除孔雀石绿，并用响应面分析了最佳吸附条件[87]。结果表明，初始溶液pH值为8.30，染料浓度为500mg/L，吸附剂用量为29.31g/L时，染料去除率为90.83%，最大吸附量为15.49mg/g。

Khan等利用废弃豌豆壳去除孔雀绿染料[88]。pH值为7时，40min内染料去除率为96%。数据与Freundlich模型吻合良好，符合准二级动力学。染料最大吸附量为14.49mg/g，吸附过程由颗粒内扩散和液膜扩散控制。

(8) 锯末

锯末本质上也是一种植物纤维废弃物，由于比较有代表性，因此单独列出。锯末含有纤维素、半纤维素、木质素和很多羟基基团，比表面积大，天然具有很好的有机物吸附能力。Khattri等使用印度楝树木屑作为吸附剂从水溶液中去除孔雀石绿[89]。染料初始浓度为12mg/L时，在pH值为7.2条件下，随着吸附剂粒度从50目减小到100目，孔雀石绿的吸附率从66.75%增加到75.78%。Song等将锯末制成生物吸附剂[65]。用三乙胺改性后，锯末的吸附率提高了632.98%。pH值为5.08时，6h内最大吸附量为697.8mg/g。吸附数据符合Freundlich模型。Garg等把锯末制备成吸附剂，并用它在间歇式反应器中去除孔雀石绿[90]。吸附剂用甲醛和硫酸处理后，染料去除率达到96%。

(9) 麦麸

麦麸是一种廉价的农业残渣。它含有多糖，如淀粉和纤维素，是一种有用的碳源。这是麦麸与其他木质纤维素材料（如松木）木屑相比的主要优势。Papinutti等使用麦麸去除孔雀石绿[91]。在28℃、pH值为7～9时，染料最大去除率为90%，最大吸附量达到240mg/g。

2.2.2 动物类

(1) 贝壳

贝壳粉是一种常用的动物来源的染料生物吸附剂，不需要任何改性活化处理，因而成本较低。通过表征实验发现，在贝壳粉的表面存在大量—OH、—CO_3和—PO_4等官能团，因此具有较强的吸附能力[92]。在 303K 和 pH 值为 8 下，在 2h 内达到吸附平衡，贝壳粉的最大单层吸附量为 42.33mg/g[92]。Shamel 等使用贝壳去除孔雀石绿[93]。在 pH 值为 10 时染料去除率为 89.48%。吸附数据遵循准二级动力学。Chowdhury 等使用海螺贝壳粉去除孔雀石绿[94]。在 303～333K 范围内，单层吸附能力随温度的升高而提高。在 303K 和 pH 值为 8 时，最大吸附量为 92.25mg/g。

(2) 羽毛

Mittal 尝试使用鸡毛去除水中的孔雀石绿[95]。从当地农场收集的鸡毛经过蒸馏，然后切成不到 1mm 长的小块。用 H_2O_2 处理后再烘干。在 pH 值为 5 时，不到 90min 吸附平衡迅速建立，染料去除率达到 85%，最大吸附量为 2.93mg/g。

(3) 蛋壳

蛋壳是一种常见的生活垃圾。Chowdhury 等使用蛋壳去除孔雀石绿[96]。在 pH 值为 9 时，90min 内达到最大吸附量。蛋壳的高吸附能力和低成本使其成为去除染料的诱人选择。

2.2.3 微生物类

微生物用于染料物理吸附多使用灭活菌体。与活细胞相比，灭活菌体的染料脱色只有吸附作用，而没有转化或降解作用。死细胞的脱色（吸附）可能是由高压灭菌过程中破裂的细胞壁面积增大所致，也可能归因于细胞壁上特殊部位的暴露[97]。

(1) 细菌

Khan 等研究了两株细菌 C1A 和 C2B 热灭活细胞对染料的脱色作用，并与活细胞进行了比较[53]。C1A 的热灭活细胞使 20mg/L 结晶紫脱色 65%，

100mg/L 孔雀石绿脱色 79%。C2B 也出现了类似的趋势。两种菌株的灭活菌体都没有胞外脱色活性。热灭活细胞的脱色表明，染料脱色 60%主要是由于吸附作用，只有近 40%的染料脱色可归因于细菌系统的降解作用。与活细胞相比，灭活菌体对染料的去除率一般较低[98]。

(2) 真菌

Singh 等使用酿酒酵母 MTCC 174 去除孔雀石绿染料[99]。从 3L 发酵罐中收集酵母细胞，洗涤后在－80℃冷冻 24h 再冻干 48h，然后加热灭活。将失活酵母细胞制成固定化细胞小珠，用于染料吸附。pH 值为 6.88 时，染料在 60min 内的最大吸附率为 96.25%。吸附数据遵循准二级动力学。Safarik 将磁性修饰的酵母细胞用作染料吸附剂[100]，为了使产品稳定且能够长时间工作，优选死酵母细胞。

Ting 等从环境中分离到一株植物内生真菌 *Diaporthe* sp. 用于三苯基甲烷染料的脱色[101]。对于棉蓝，无论是活细胞还是死细胞，脱色率都达到 85%。而对于结晶紫、孔雀石绿和甲基紫，活细胞脱色率都明显高于死细胞。死细胞的优势是脱色迅速，1～3h 可达到最大脱色活性，而活细胞脱色需要 24h 以上才能达到最大脱色率。对于不同类型的染料，死细胞的吸附效果也有不同。Przystas 发现大多数真菌死细胞对灿烂绿的 24h 吸附量在 42%～57%之间，时间延长到 120h 后达到 77%，但对伊文思蓝的 24h 吸附量为 2.9%～23.3%，在 120h 后仅有 8.1%～31%[98]。简青霉的活细胞在更短时间内（0.5～2h）对结晶紫、甲基紫、孔雀石绿和棉蓝的脱色率（75.6%～90.8%）都明显高于死细胞（43.9%～75.2%，4h）[102]。

(3) 藻类

从藻类废料中提取的小球藻可用作低成本生物吸附剂。Tsai 等利用死亡微藻快速去除了孔雀石绿[103]。与市售活性炭进行比较，藻类可以作为一种低成本的生物吸附剂有效地去除水溶液中的三苯基甲烷染料。

2.3 微生物转化或降解

由于三苯基甲烷染料具有广泛的应用和富集能力，在废水中普遍存在，会污染水环境。为了去除水体中的合成染料，已经开发了一系列物理、化学和生物方法。而生物脱色包括生物物理吸附和微生物转化或降解两种方式。生物物理吸附在第 2.2 节中已有详细阐述，本小节将主要讨论三苯基甲烷染料的微生物转化或降解。废水中三苯基甲烷染料的生物降解是一种非常有效且环境友好的方法[55,104]。

染料的生物降解可通过胞外和胞内酶介导。在细胞内生物降解的情况下，生物积累是脱色的主要机制。最近的研究结果还表明，在三苯基甲烷染料的去除过程中，非酶的、耐高温的低分子量化合物也可能参与其中[105-107]。此外，在类 Fenton 反应过程中，白腐菌和褐腐菌产生的羟基自由基被认为是影响三苯基甲烷染料脱色的潜在因素[108,109]。

微生物脱色中常用的三苯基甲烷染料如表 2-1 所示。

表 2-1　微生物脱色中常用的三苯基甲烷染料

染料	最大光吸收/nm	参考文献
酸性品红	544	Ren 等[110], Casas 等[111], El-Sheekh 等[112], Jing 等[113]
灿烂绿	623	Ren 等[110], Casas 等[111], El-Sheekh 等[112], Jing 等[113], Kumar 等[114]
溴酚蓝	590	Ren 等[110], Gomaa 等[115], Saparrat 等[116], Casas 等[111], Kumar 等[114], Przystas 等[98]
棉蓝	620	Kalyani 等[117], Parshetti 等[118], Shedbalkar 等[119]
结晶紫	530	Jang 等[120], Šafaříková 等[121], Eichlerová 等[122], Kalyani 等[117], Chen 等[123,124], Safarik 等[125], Abedin[126], Youssef 等[127], Moturi 等[128], Yan 等[106], Ayed 等[129], Ghasemzadeh 等[130], Pandey 等[131], Parshetti 等[132], Torres 等[133], Yang 等[134,135], Cheriaa 等[136]
孔雀石绿	615	Jang 等[120], Levin 等[137], Oranusi 等[138], Eichlerová 等[122], Kalyani[117], Papinutti 等[91], Parshetti 等[132], Ren 等[110], Saratale 等[118], Daneshvar 等[139], Abedin[126], Deng 等[140], Youssef 等[127], Ayed 等[129], Jing 等[113], Moturi 等[128], Chen 等[123], Du 等[141], Shedbalkar 等[142], Vasdev 等[143], Yang 等[134,135], Cheriaa 等[136], Jasińska 等[57,58], Karimi 等[108], Lv 等[144], Mukherjee 等[145], Yan 等[106]

2.3.1　细菌

三苯基甲烷染料很容易通过传统的废水处理过程，并对周围的生物群造成严重的毒性[129]。然而，研究人员已经确定了许多细菌种类，可以在适当的培养条件下对染料进行脱色和解毒。在这方面，Sani 等研究了 *Kurthia* sp. 对合成、纺织和染料废水（品红、结晶紫、孔雀石绿、副玫瑰苯胺和灿烂绿）的脱色[146]。

研究人员发现在有氧条件下，细菌对 10μmol/L 品红、结晶紫、孔雀石绿、副玫瑰苯胺和灿烂绿染料的脱色率分别为 92%、96%、96%、100%和 100%。与真实的纺织品和染料废水相比，合成废水的脱色程度更高（56%）。后来，Parshetti 等研究了在静态缺氧条件下，*Kocuria rosea* MTCC 1532 在添加糖蜜的半合成介质中对孔雀绿染料的脱色[118]。作者报道，该菌株可通过还原酶催化的代谢反应将染料（初始浓度为 50mg/L）完全降解成无毒化合物。在另一项研究中，Parshetti 等调查了 *Agrobacterium radiobacter* MTCC 8161 在静态缺氧条件下对结晶紫的生物降解潜力[132]。在初始浓度为 10mg/L 的染料完全降解过程中，发现漆酶和氨基比林 *N*-脱甲基酶的代谢产物。

Chen 等从炼油工业的冷却系统中分离出 *Shewanella* sp. NTOU1，用于在优化的静态厌氧条件下生物降解结晶紫染料[124]。研究人员发现，通过将染料生物转化为 *N*,*N*-二甲基氨基苯酚和米氏酮（MK），该菌株可以显著地使染料脱色高达 1500mg/L。此外，生物转化的化合物被分解成无毒的产物。反应介质中添加了甲酸盐和柠檬酸铁作为电子供体，提高了细菌的活性。因此，基于所获得的结果，作者报道了该菌株能够以比其他先前报道的微生物更具结晶紫降解能力。Ayed 等研究了从纺织废水中分离的 *Staphylococcus epidermidis* 在振荡条件下的生物降解和脱色能力[147]。研究人员观察到，该细菌能在 14h 内分别显著去除初始浓度为 750mg/L 的结晶紫、酚红、孔雀石绿、甲基绿和品红染料的 62%、80%、33%、80%和 37%。此外，降解的最终产物是对小麦发芽无毒的化合物。后来，Ogugbue 等从纺织废水中分离出嗜水气单胞菌，并将其用于含有 50mg/L 染料（碱性紫-14、碱性紫-3 和酸性蓝-90）废水的脱色[148]。研究人员报告说，在优化的有氧条件下培养 24h，这种细菌可以去除 72%～96%的染料。基于对处理过的废水的生化分析，发现有毒染料转化为无毒代谢物而发生脱色。在另一项研究中，*Pseudomonas putida* MTCC 4910 被用于在厌氧培养基中生物降解碱性紫-3 和酸性蓝-93[149]。他们发现该菌株能在 5h 内完全还原降解浓度为 50mg/L 的染料并生成多种化合物。

相关研究汇总于表 2-2 中。

表 2-2 三苯基甲烷染料细菌生物脱色的研究[150]

染料	细菌种类	去除率/%	参考文献
孔雀石绿	*Pandoraea pulmonicola* YC32	85.2	Chen 等[151]
碱性紫-14、碱性紫-3 和酸性蓝-90	*Aeromonas hydrophila*	72～96	Ogugbue 等[148]
结晶紫	*Shewanella* sp. NTOU1	100	Chen 等[124]
结晶紫、酚红、孔雀石绿、甲基绿和品红	*Staphylococcus epidermidis*	33～80	Ayed 等[147]

续表

染料	细菌种类	去除率/%	参考文献
品红、结晶紫、孔雀石绿、副玫瑰苯胺和灿烂绿	*Kurthia* sp.	90～100	Sani 等[146]
孔雀石绿和灿烂绿	*Pseudomonas otitidis* WL-13	95	Jing 等[113]
苯胺蓝	*Shewanella oneidensis* MR-1	90	Wu 等[152]
碱性紫-3 和酸性蓝-93	*Pseudomonas putida* MTCC 4910	100	Khan 等[149]
孔雀石绿	*Kocuria rosea* MTCC 1532	100	Parshetti 等[118]
结晶紫	*Agrobacterium radiobacter* MTCC 8161	100	Parshetti 等[132]

此外，一些菌株能够降解甚至完全矿化合成染料[153,154]。与真菌相比，细菌通常更容易培养，生长更快，更易于操作。这些特性使细菌成为研究三苯基甲烷染料生物降解的理想对象。假单胞菌属、芽孢杆菌属、柠檬酸杆菌属、脱硫弧菌属、诺卡氏菌属和分枝杆菌属的成员通常被认为是良好的三苯基甲烷染料脱色剂[138,140,155]。然而，近年来，陆续有新脱色菌种被报道，如希瓦氏菌、醋酸钙不动杆菌、嗜水气单胞菌、木糖氧化无色杆菌、鞘氨醇单胞菌，以及耐辐射球菌、肠杆菌和表皮葡萄球菌等[107,110,129,144,153,156-158]。在上述研究中检测到的细菌大多是从被染料污染的环境中分离出来的，通常使用单一菌株对三苯基甲烷染料进行脱色。

2.3.2 真菌

研究最广泛的三苯基甲烷染料脱色真菌是属于担子菌纲的白腐真菌[137]。*Phanerochaete chrysosporium* 菌株是合成染料降解研究中最常用的担子菌模型。该属的成员也被认为是各种三苯基甲烷染料的良好脱色剂，例如结晶紫、副玫瑰苯胺、甲酚红、溴酚蓝、乙基紫、孔雀石绿和灿烂绿[115,105]。其他担子菌菌种也被发现可以去除三苯基甲烷染料。例如，*Dichomitus squalens* CCBAS 750 被证明是几种不同化学合成染料的良好脱色剂，包括孔雀石绿和结晶紫[122]。Casas 等描述了 *Trametes versicolor* ATCC 42530 对酸性品红、灿烂绿 1、碱性品红、甲基绿和酸性绿 16 的有效脱色[111]。Saparrat 等报告称，来自美洲和东非温带和热带地区的白腐真菌 *Grammothele subargentea* LPSC 436 能够有效用于偶氮、杂环和三苯基甲烷染料的脱色[116]。尽管三苯基甲烷染料对生长有强烈的抑制作用，但该菌株能高效地降解灿烂绿、结晶紫和品红。Vasdev 的研究发现，结晶紫、溴酚蓝和孔雀石绿的有效脱色是由六种从自然界分离的白腐真菌实现的[143]。

此外，据报道，子囊菌具有生长迅速、生物量大、细胞比表面积大的特点，可与环境形成牢固的物理接触，可用于去除各种异生素，包括三苯基甲烷染料[54,159]，其中许多来源于被染料污染的环境（土壤、废水和污泥）。例如，来源于海水、海洋沉积物和海草的真菌，*Phialophora* sp.、*Penicillium* sp. 和 *Cladosporium* sp. 被证明可以完全使结晶紫脱色[133]。从水样中分离出的 *Aspergillus* sp. CB-TKL-1 能够使几种结构不同的染料快速脱色，尤其是甲基紫和灿烂绿[114]。*Aspergillus ochraceus* NCIM-1146 可分别将孔雀石绿、棉蓝、结晶紫和甲基紫脱色 98%、92%、61%和 57%。该过程涉及微生物降解代谢，而不仅仅是生物吸附[118]。从含有染料的废水中分离出的 *Fusarium solani* 能够通过生物吸附高效地使孔雀石绿和结晶紫脱色，然后在细胞内降解为无色代谢物[126]（Abedin2008）。从纺织印染厂周围的土壤中分离出来的 *Penicillium pinophilum* IM6480 和 *Myrothecium roridum* IM 6482 也存在类似的孔雀石绿脱色方式[8]。

Kwasniewska 报道了氧化红酵母，如 *Rhodotorula* sp. 和 *Rhodotorula rubra*，能够降解液体肉汤中的结晶紫[160]。生长培养基中测试的结晶紫的初始浓度为 10mg/L，孵育四天后，上清液在 600nm 处的吸光度变得不可测量，表明两种氧化红酵母降解了结晶紫。为了测试细胞对染料的吸附性，将生物降解后的细胞在 70%（体积分数）乙醇中超声处理。提取物在 600nm 处没有显示任何光吸收。还观察到，即使降解期长达 30 天，酿酒酵母也不会降解液体培养基中的结晶紫。由于染料吸附到细胞上，光密度几乎没有降低。酿酒酵母不能降解结晶紫显然不是由于染料的任何毒性作用。相差显微观察发现酵母在对照瓶和测试瓶中均生长良好。在培养的第二天到第四天之间，两种氧化红酵母对结晶紫产生了线性降解，这表明存在降解结晶紫的酶系统。Moe 等从缅甸和泰国的养鱼场和传统发酵渔业产品中，筛选出一株耐盐酵母菌 *Debaryomyces nepalensis*[161]。染料降解实验中，孔雀石绿浓度逐渐降低，但隐性孔雀石绿含量增加直到孔雀石绿初始含量的 40%左右。实验结束时，核磁共振没有检测到孔雀石绿以及它的同系物或其他芳香代谢产物。结果表明 *Debaryomyces nepalensis* 将孔雀石绿转化为大约 40%的隐性孔雀石绿和 60%的其他代谢物。

2.3.3 放线菌

Yatome 等发表了放线菌对三苯基甲烷染料生物降解的第一篇文章[162]。用 *Nocardia corallina* 和 *Nocardia globerula* 两种放线菌对结晶紫进行脱色。结果表明表明放线菌的脱色是细胞内的，因为培养滤液没有活性。这些染料在 24h 内完全脱色。它们还检测到结晶紫的降解产物是MK。该作者的另一篇文章中，*N. corallina* 能将98%的结晶紫降解为MK，但不能使金胺O染料脱色[163]。与之相比，甲基紫、乙基紫、碱性品红和维多利亚蓝的脱色程度要弱得多。脱色率取决于培养基中降解菌的初始浓度和细胞的生长阶段。结晶紫的脱色仅在低浓度（5μmol/L）下观察到，而在7μmol/L时细胞生长完全被抑制。培养基类型对脱色率有影响，结晶紫在LB、Bennett和基础培养基中的脱色率每分钟分别为10.6nmol/mg、6.9nmol/mg和2.8nmol/mg。*N. corallina* 细胞匀浆和胞外提取液对结晶紫没有脱色作用。*N. corallina* 细胞洗涤后在缓冲液中重悬时，未观察到脱色活性，但在LB培养基中孵育后，活性恢复。生物降解产物依然为MK。

2.3.4 藻类

Rajeshkannan 等尝试了用水生植物黑藻从溶液中去除孔雀石绿[164]。pH值为8.0时的吸附量为91.97mg/g。Kumar等提出了一种使用淡水藻类 *Pithophora* sp. 去除孔雀石绿的方法[165]。在pH值为6、30℃条件下，染料的最大吸附量为42.2mg/g，去除率达到98.88%。Jerold等使用褐藻去除孔雀石绿染料[166]。在pH值为10时达到最大的染料吸附。数据遵循二级动力学，与Langmuir模型拟合良好，膜扩散是控速步骤。最近，Han等评估了生活在淡水中的光自养蓝藻 *Synechococcus elongatus* PCC 7942对孔雀绿的脱色能力[167]。结果表明，通过生物吸附和生物富集，合成球菌对染料的去除率为99.5%，但未对染料进行降解或化学修饰。接下来，作者建立了一个工程化的 *Synechococcus* 菌株来降解摄取后的染料。三苯基甲烷还原酶基因 *katmr* 是异源表达的，可高效表达可溶性重组蛋白。该工程菌在低细胞密度和逆境条件下表现出较强的脱色能力，可以将孔雀石绿降解为小分子4-甲基氨基苯甲酸和4-羟基苯胺。经蓝藻工程菌处理后，在孔雀石绿存在下，小麦种子的生长完全恢复。这些结果表明，工程菌作为光合细胞工厂在去除废水中孔雀石绿方面具有潜在的应用前景。

2.3.5 混合菌群

混合菌群在处理印染废水方面也非常有效。Cheriaa等报道，与单一菌种相比，含有放射杆菌、芽孢杆菌、鞘氨醇单胞菌和亲水曲霉菌株的混合菌群对于孔雀石绿和结晶紫的脱色更快更有效[136]。类似地，由芽孢杆菌、产碱杆菌和气单胞菌组成的菌群比单一菌种对酸性紫罗兰17的脱色效果更好[168]。由于微生物之间的协同作用，微生物菌群可以提高染料的脱色率。

Idaka 等在染料驯化的活性污泥中，研究了刚果红、酸性橙和结晶紫的趋归。他们观察到，随着染料浓度的增大，脱色率降低[169]。Ogawa 等报道了用未驯化的污泥研究各种染料对其生长和呼吸的抑制作用[170]。作者还使用了适应不同三苯基甲烷染料的污泥，这些污泥对不同浓度的染料具有适应性[171]。观察到甲基紫对未驯化微生物的抑制水平随着染料浓度的增大而增强。与其他染料相比，甲基紫显示出更强的抑制作用。驯化的微生物在低染料浓度下表现出负抑制，在高浓度下表现出正抑制。该团队还报道了某些碱性染料对微生物生长的抑制作用[172]。他们使用枯草芽孢杆菌作为测试生物，并观察到随着染料的加入，对数期细胞群体的平均生长率和稳定期的细胞浓度都降低了。还报道了三苯基甲烷染料（结晶紫和甲基紫）强烈抑制细胞生长。染料对细胞化学成分的抑制机理有待进一步阐明。为此，他们测定了对数期和稳定期收获细胞的核酸含量，发现核糖核酸含量随着甲基紫浓度的增大而减少。这种趋势在对数期比稳定期更明显。核酸含量比，即核糖核酸/脱氧核糖核酸，随着甲基紫浓度的增大而降低，因此他们得出结论，染料更倾向于抑制蛋白质合成，而不是抑制细胞分裂。由于抑制作用，细胞形态也发生了变化。

2.4 微生物脱色的影响因素

2.4.1 碳源

三苯基甲烷染料的生物脱色受到许多操作参数的强烈影响。由于对生物生长和代谢有巨大影响，培养基成分（特别是碳、氮化合物）会强烈影响合成染料的脱色。葡萄糖和蔗糖似乎是微生物降解诸如碱性紫 3 和孔雀石绿的最合适的碳源[173,174]。在含有葡萄糖的培养基中加速去除染料，通常是由于生物量的增长。然而，由于葡萄糖的高成本，各种廉价的化合物被用作脱色过程中的碳源[59]。在四种不同的碳源（葡萄糖、蔗糖、淀粉和柠檬酸钠）中，淀粉更有利于 *F. solani* 菌株（Abedin 2008）对结晶紫的脱色[126]。还有 Parshetti 等成功应用糖蜜（代替葡萄糖或蔗糖）促进 *K. rosea* MTCC 1532 对孔雀石绿进行脱色[118]。

2.4.2 氮源

氮源也可影响染料的生物降解，因为它们影响木质素降解酶如漆酶和过氧化物酶的合成。一般来说，真菌木质素降解酶的产生通常是由缺氮引起的。例如，Ghasemzadeh 等报告称，在氮源限制条件下，*P. chrysosporium* 产生大量与结晶紫有效消除相关的 LiP 和 MnP[130]。另外，染料脱色还受氮源类型的影响。据推测，与无机氮源相比，有机化合物，如蛋白胨或酵母提取物，经常加速染料脱色过程。事实上，酵母提取物被认为是各种三苯基甲烷染料高效脱色的最佳氮源[132,136]。酵母提取物被认为是必要的培养基补充物，这可能与其作为三苯基甲烷染料酶促还原电子供体的 NADH 再生的适宜性有关。作为一种重要的培养基补充剂，酵母提取物充当三苯基甲烷染料酶促还原的电子供体，其识别可能与其对 NADH 再生的适应性有关[120]。

2.4.3 pH

脱色环境的 pH 值影响染料分子的跨细胞膜运输以及木质素分解酶对三苯基甲烷染料的细胞外生物转化。Nozaki 等在研究担子菌属 21 种真菌对三苯基甲烷染料的脱色时发现，最佳 pH 值为 3～5[175]。随后，该团队又发现一株对溴酚蓝去除率最高的香菇菌株，最佳脱色 pH 值是 4[176]。通常，真菌和酵母更喜欢酸性 pH 值。然而，孔雀石绿在 pH 值为 7 时被曲霉 *P. ochrochloron* 菌丝体有效脱色[142]。类似地，Deivasigamani 等报道了在 pH 值为 7 下培养的克鲁氏梭菌 *C. krusei* 将碱性紫 3 完全脱色[174]。Yang 等估计，pH 值为 6 的生长培养基加速了青霉菌株 YW01 对孔雀石绿的脱色。细菌通常会在中性和碱性 pH 下使染料脱色[134]。Du 等观察到，假单胞菌 DY1 对孔雀石绿的脱色的适宜 pH 值在 5.5～8 之间，最高脱色率在 pH 值为 6.6[141]。在 pH 值为 5～7 的范围内，钙醋酸菌 YC210 对维多利亚蓝 R 的脱色效果最好[153]。铜绿假单胞菌 BCH 在 30min 内将酸性紫罗兰 19 脱色至 98%，最适 pH 值为 7[177]。

2.4.4 温度

在大多数情况下，随着温度的升高呼吸作用和底物新陈代谢加快，三苯基甲烷染料的生物去除率会增大，更适合细胞生长。值得一提的是，一些三苯基甲烷染料降解酶即使在高温下也是有活性的和稳定的。例如，孔雀石绿在30～80℃的温度范围内被 *Trametes trogii* Berk S0301 有效脱色，最佳脱色温度为50～70℃[178]。根据 Zhang 等的研究，在25～90℃范围内，*Bacillus vallismortis* fmb-103 的孢子漆酶非常稳定[179]。NADH 依赖性的柠檬酸杆菌的三苯基甲烷染料还原酶 KCTC 催化结晶紫生成无色染料，在60℃时显示出最强活性[120]。

2.4.5 溶解氧

三苯基甲烷染料的生物降解在有氧和无氧条件下都可能发生，具体取决于特定微生物的代谢特征。大多数真菌都是专性需氧菌，它们需要氧气才能生长和维持活性。为了满足其氧气需求并增强氧气-液体的质量传递，需要进行搅拌。大多数文献数据表明，真菌在摇动条件下能更快地去除三苯基甲烷染料[98,118]。通常，在摇动条件下，细菌降解三苯基甲烷染料的能力也得到改善[54,141]。相比之下，蜡状芽孢杆菌 DC11 对孔雀石绿的脱色需要厌氧或微需氧条件，并且在需氧培养中受到强烈抑制[140]。这可能是因为溶解氧经常抑制还原酶所驱动的脱色反应。在有氧条件下，氧气可以与染料竞争还原的电子载体[180]。

2.4.6 染料浓度

通常，脱色效率随着三苯基甲烷染料浓度的增大而降低，这主要是由于染料的毒性增强。当微生物生长减慢时，酶的分泌也会受到限制，尤其是在固体培养基中[118,122]。然而，染料的初始浓度具有很强的驱动力，可克服染料和细胞之间的传质阻力。例如，随着染料浓度增大至450mg/L，*A. calcoaceticus* 对维多利亚蓝 R 的脱色作用增强[153]。同样，*P. chrysosporium* ATCC 34541 对维多利亚蓝的脱色被证明在50～350mg/L 的范围内几乎与染料浓度无关[115]。

2.5 脱色酶

各种细菌和真菌对合成染料的解毒作用通常由氧化还原酶介导。木质素过氧化物酶、锰过氧化物酶和漆酶是三苯基甲烷染料脱色的主要氧化还原酶[181]。

2.5.1 氧化酶

漆酶是含铜的酶，可催化酚类化合物的氧化，同时将一个双氧分子还原为两个水分子。漆酶可以在氧化还原介质的存在下起作用，例如 2,2′-联氮双(3-乙基苯并噻唑啉-6-磺酸)二铵盐（ABTS）、1-羟基苯并三唑（HBT）和 3-羟基邻氨基苯甲酸，分解包括合成染料在内的几种难以降解的化合物[182]。基于实验证据，在图 2-1 中提出了漆酶介导的结晶紫染料脱色机理[132]。锰过氧化物酶和木质素过氧化物酶也被证明参与了多种芳香烃和酚类化合物的降解，包括合成染料[134,183]。通常，过氧化物酶是在 H_2O_2 存在下催化血红素蛋白的转化。而锰过氧化物酶催化酚类化合物的转化，这与 Mn(Ⅱ) 到 Mn(Ⅲ) 的氧化有关[184]。类似地，木质素过氧化物酶进行酚类化合物的单电子氧化，但通常也进行与其他过氧化物酶不相关的反应，即非酚类芳香族底物的氧化[185]。木质素过氧化物酶

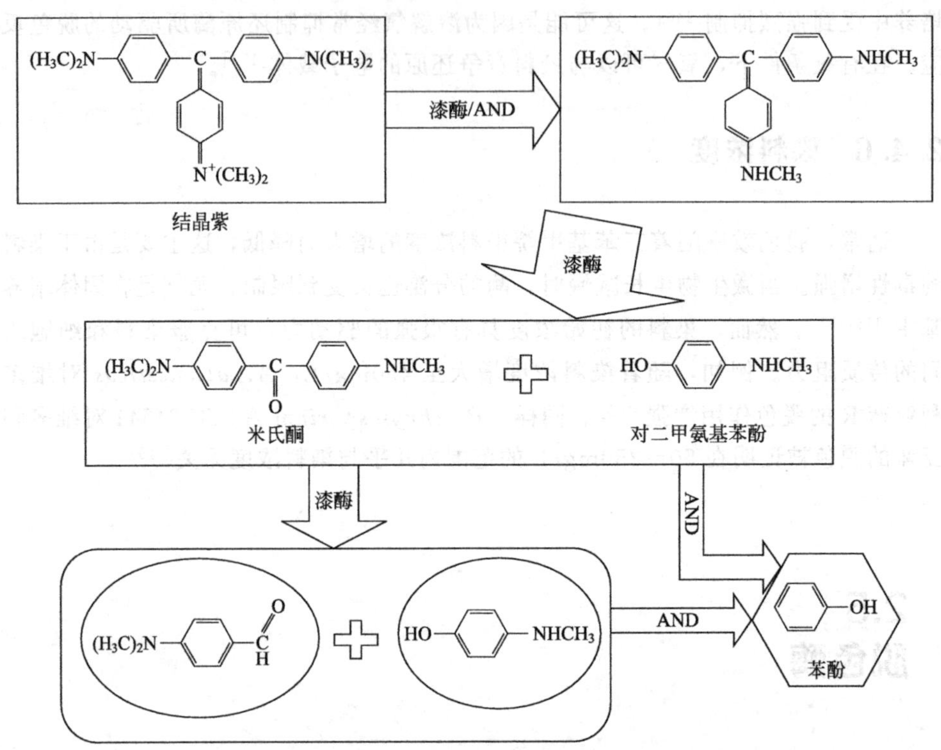

图 2-1 漆酶介导的结晶紫染料脱色机理[132]

AND：氨基比林-*N*-去甲基化酶

的脱色活性可以在某些化合物（例如藜芦醇）的存在下增强，藜芦醇充当木质素过氧化物酶和染料之间的氧化还原中间体[186]。根据Gomaa等获得的数据，漆酶、加氧酶/氧化酶和/或热稳定的非酶因子被认为是*P. chrysosporium* ATCC 34541降解维多利亚蓝的最可能因素[115]。漆酶和锰过氧化物酶的活性较高，与*Pleurotus pulmonarius*（FR）的乙基紫脱色有关[187]。据报道，一种白腐真菌*Ischnoderma resinosum*能够在液体培养基中培养的20天之内，通过漆酶和锰过氧化物酶的作用，使孔雀石绿和结晶紫几乎完全脱色[188]。Jasińska等也描述了在*M. roridum* IM 6482的漆酶培养物中，油菜籽压榨饼中孔雀石绿的降解[58]。*Coprinus comatus*的漆酶可对90%以上的孔雀石绿进行脱色，在氧化还原介质（尤其是1-羟基苯并三唑）的存在下，脱色率更高[189]。*Lentinula edodes* CCB-42菌株能降解甲基紫、乙基紫和甲基绿[190]。由于该过程受Mn离子和H_2O_2的强烈影响，因此锰过氧化物酶的作用被认为是消除染料的主要机理。Shedbalkar等报道了由*Penicillium ochrochloron* MTCC517木质素过氧化物酶诱导的棉蓝生物转化[119]。同样的菌株也被发现通过过氧化物酶介导的反应来降解孔雀石绿。Balan等研究了以香蕉皮为固体基质生产的漆酶对孔雀石绿的去除作用[191]。当漆酶和1-羟基苯并三唑的浓度分别为2.16U/mL和0.85mmol/L时，3h后95.8mg/L孔雀石绿的去除率约为96%。Zhuo等成功将*Ganoderma* sp. En3的漆酶基因lac-En3-1在重组毕赤酵母中表达[192]。纯化的rLAC-EN3-1漆酶在碱性条件下表现出很强的稳定性，例如对十二烷基硫酸钠（SDS）、EDTA等酶抑制剂也表现出较强的抗性；能耐受很高浓度的$CuSO_4$、$MnSO_4$和$MgSO_4$（100mmol/L）；对工业上广泛使用的甘油、乙二醇、丁二醇等有机溶剂具有很强的耐受能力。脱色实验结果表明，rLAC-EN3-1对孔雀石绿具有较强的脱色能力。脱色后产物对平菇的毒性比原始染料低得多。

微生物去除三苯基甲烷染料也可能是膜相关氧化还原酶作用的结果。例如，*Mycobacterium avium* A5的膜部分具有比粗提取物高约5倍的孔雀石绿特异性脱色率，这表明膜相关蛋白质，如细胞色素P-450参与其中[193]。在*Cunninghamella elegans* ATCC 36112培养物中也报道了细胞色素P-450单加氧酶参与孔雀石绿的脱色[194]。

2.5.2 还原酶

在三苯基甲烷还原酶的控制下，三苯基甲烷染料也可以通过依赖于NADH/NADPH的还原进行脱色。从*Citrobacter* sp. KCTC 18061P分离出第一个纯化并经过生物学鉴定的三苯基甲烷还原酶[120]。然而，在 *Pseudomonas aeruginosa* NCIM 2074、*Exiguobacterium* sp. MG2 和 *Mucor mucedo* 细胞中也鉴定出了三苯基甲烷还原酶[195-197]。结果表明，三苯基甲烷还原酶的脱色活性与染料的化学结构有关。最有效的三苯基甲烷还原酶底物似乎是孔雀石绿，而结晶紫则效果较差，可能是因为增加了一个二甲氨基[120]（Jang 等，2005 年）。Kim 等认为，其他结构的三苯基甲烷染料，如灿烂绿、溴酚蓝、甲基红和刚果红，与三苯基甲烷还原酶底物结合口袋的大小和疏水性不相容[198]。基于结构检测，提出了模型化的三元蛋白质/辅因子/底物复合物的结构以及三苯基甲烷还原酶对孔雀石绿脱色的机理（图 2-2）。

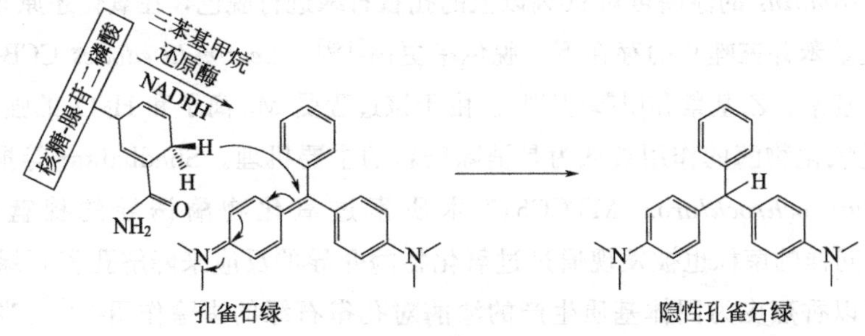

图 2-2　三苯基甲烷还原酶对孔雀石绿脱色的机理[198]

使用真菌和细菌对合成染料进行生物降解已成为处理染料废水的一种有前景的方法。如今，分子生物学提供了各种工具来强化生物体的自身能力并优化降解途径。开发主要用于生物修复的基因工程微生物的主要方法包括：改变微生物酶的特异性和亲和力以及对生物过程的监测和调节。此外，鉴定了多种使三苯基甲烷染料脱色的酶，并将其在多种宿主生物中表达[110,120,195,199]。在一株 *Amycolatopsis sp*. 中，Chengalroyen 等鉴定了四个三苯基甲烷染料生物降解基因，编码 3-脱氧-7-磷酸庚酮酸合酶、*N*5,*N*10-亚甲基四氢甲烷喋呤还原酶、多囊肾结构域Ⅰ和葡萄糖/山梨糖脱氢酶[200]。这些基因在 *Streptomyces lividans* TK23 中的协同表达作用导致结晶紫完全脱色。在 *Mycobacterium* sp. mc2 155 和三株 *Rhodococcus* sp. 菌株中也测试了它们的活性，两个菌株的脱色染料范围都有所扩大，表明克隆的基因在宿主体内具有新的功能。此外，Fu 等提出了一种基于柠檬酸杆菌三苯基甲烷还原酶过度表达的拟南芥植物修复系统[201]。转基因拟南芥植株的形态和生长表现出对结晶紫和孔雀石绿的耐受性，以及对这些染料脱色的能力显著增强。

三苯基甲烷染料生产规模大、应用范围广、性质稳定，但其严重的健康风险也造成了巨大的环境危害。迄今为止，已开发出许多用于处理三苯基甲烷染料废水的策略。其中，微生物脱色被视为最具前途的染料修复方法。可用于三苯基甲烷染料微生物脱色的微生物有很多，包括细菌、藻类和真菌等。除生物吸附外，染料的脱色主要通过细胞外和/或细胞内酶（例如漆酶、过氧化物酶和还原酶）的作用实现。生物处理是最终控制纺织和印染行业污染的唯一途径。因此，需要越来越多的研究和开发工作，如新菌种筛选、新脱色酶发现以及新工艺建立等，来推动染料废水处理的可持续发展。

参 考 文 献

[1] Pan T，Ren S，Xu M，Sun G，Guo J. Extractive biodecolorization of triphenylmethane dyes in cloud point system by *Aeromonas hydrophila* DN322p. Applied Miorobiology and Biotechnology，2013，97 (13)：6051-6055.

[2] Watharkar A D，Khandare R V，Kamble A A，Mulla A Y，Govindwar SP，Jadhav JP. Phytoremediation potential of *Petunia grandiflora* Juss.，an ornamental plant to degrade a disperse，disulfonated triphenylmethane textile dye Brilliant Blue G. Environmental Science and Pollution Research，2013，20 (2)：939-949.

[3] Zhou X J，Guo W Q，Yang S S，Zheng H S，Ren N Q. Ultrasonic-assisted ozone oxidation process of triphenylmethane dye degradation：Evidence for the promotion effects of ultrasonic on malachite green decolorization and degradation mechanism. Bioresource Technology，2013，128：827-830.

[4] Kumar M A，Harthy D K，Kumar V V，Balashri K G，Seenuvasan M，Anuradha D，Sivanesan S. Detoxification of a triphenylmethane textile colorant using acclimated cells of bacillus mannanilyticus strain AVS. Environmental Progress and Sustainable Energy，2017，36 (2)：394-403.

[5] Chen S H，Cheow Y L，Ng S L，Ting A S Y. Biodegradation of triphenylmethane dyes by non-white rot fungus penicillium simplicissimum：enzymatic and toxicity studies. International Journal of Environmental Research，2019，13 (2)：273-282.

[6] Ong S T，Keng P S，Lee W N，Ha S T，Hung Y T. Dye waste treatment. Water，2011，3 (1)：157-176.

[7] Holkar C R，Jadhav A J，Pinjari D V，Mahamuni N M，Pandit A B. A critical review on textile wastewater treatments：Possible approaches. Journal of Environmental Management，2016，182：351-366.

[8] Jasińska A，Paraszkiewicz K，Słaba M，Długoński J. Microbial decolorization of triphenylmethane dyes//Singh S N. Microbial degradation of synthetic dyes in wastewaters. Environmental Science and Engineering. Berlin：Springer，2015：169-186.

[9] Forgacs E，Cserhati T，Oros G. Removal of synthetic dyes from wastewaters：a review. Environment international，2004，30 (7)：953-971.

[10] Joshi M，Purwar R. Developments in new processes for colour removal from effluent. Review of Progress in Coloration and Related Topics，2004，34 (1)：58-71.

[11] Roy D, Wang G, Adrian D D. A simplified solution technique for carbon adsorption model. Water Research, 1993, 27 (6): 1033-1040.

[12] Shah I, Adnan R, Wan Ngah W S, Mohamed N. Iron impregnated activated carbon as an efficient adsorbent for the removal of methylene blue: regeneration and kinetics studies. Plos One, 2015, 10 (4): e0122603.

[13] Burca S, Indolean C, Maicaneanu A. Malachite green dye adsorption from model aqueous solutions using corn cob activated carbon (ccac). Studia Universitatis Babes-Bolyai Chemia, 2017, 62 (4): 293-307.

[14] Ravindiran G, Ganapathy G P, Josephraj J, Alagumalai A. A critical insight into biomass derived biosorbent for bioremediation of dyes. Chemistryselect, 2019, 4 (33): 9762-9775.

[15] Bhatnagar A, Sillanpaa M, Witek-Krowiak A. Agricultural waste peels as versatile biomass for water purification—A review. Chemical Engineering Journal, 2015, 270: 244-271.

[16] Bizuneh A. Textile effluent treatment and decolorization techniques. Bulgarian Journal of Science Education, 2012, 21: 434-456.

[17] Crini G. Non-conventional low-cost adsorbents for dye removal: A review. Bioresource Technology, 2006, 97 (9): 1061-1085.

[18] Anjaneyulu Y, Sreedhara Chary N, Samuel Suman Raj D. Decolourization of industrial effluents—available methods and emerging technologies—A Review. Reviews in Environmental Science and Bio/Technology, 2005, 4 (4): 245-273.

[19] Mishra G, Tripathy M. A critical review of the treatments for decolourization of textile effluent. Colourage, 1993, 40: 35-35.

[20] Meyers R A. Environmental analysis and remediation. New York: John Wiley, 1998.

[21] Vijay E V V, Jerold M, Sankar M S R, Lakshmanan S, Sivasubramanian V. Electrocoagulation using commercial grade aluminium electrode for the removal of crystal violet from aqueous solution. Water Science and Technology, 2019, 79 (4): 597-606.

[22] Randall S, Harper D, Brierley B. Ecological and ecophysiological impacts of ferric dosing in reservoirs. Hydrobiologia, 1999, 395: 355-364.

[23] Dalzell D J B, Macfarlane N A A. The toxicity of iron to brown trout and effects on the gills: a comparison of two grades of iron sulphate. Journal of Fish Biology, 1999, 55 (2): 301-315.

[24] Petzold G, Schwarz S. Dye removal from solutions and sludges by using polyelectrolytes and polyelectrolyte-surfactant complexes. Separation and purification technology, 2006, 51 (3): 318-324.

[25] van Anholt R D, Spanings F A T, Knol A H, van der Velden J A, Wendelaar Bonga S E. Effects of iron sulfate dosage on the water flea (*Daphnia magna* Straus) and early development of Carp (*Cyprinus carpio* L.). Archives of Environmental Contamination and Toxicology, 2002, 42 (2): 182-192.

[26] Koyuncu I. Reactive dye removal in dye/salt mixtures by nanofiltration membranes containing vinylsulphone dyes: effects of feed concentration and cross flow velocity. Desalination, 2002, 143 (3):

243-253.

[27] Sen S, Demirer G N. Anaerobic treatment of synthetic textile wastewater containing a reactive azo dye. Journal of Environmental Engineering, 2003, 129 (7): 595-601.

[28] Xu Y, Lebrun R E, Gallo P J, Blond P. Treatment of textile dye plant effluent by nanofiltration membrane. Separation Science and Technology, 1999, 34 (13): 2501-2519.

[29] Vishnu G, Joseph K. Nanofiltration and ozonation for decolorisation and salt recovery from reactive dyebath. Coloration Technology, 2007, 123 (4): 260-266.

[30] Craft T F, Eichholz G G. Decoloration of textile dye waste solutions by combined irradiation and chemical oxidation. Nuclear Technology, 1973, 18 (1): 46-54.

[31] Behnajady M A, Modirshahla N, Shokri M, Vahid B. Effect of operational parameters on degradation of Malachite Green by ultrasonic irradiation. Ultrasonics Sonochemistry, 2008, 15 (6): 1009-1014.

[32] Han R, Wang Y, Sun Q, Wang L, Song J, He X, Dou C. Malachite green adsorption onto natural zeolite and reuse by microwave irradiation. Journal of Hazardous Materials, 2010, 175 (1-3): 1056-1061.

[33] Gogate P R, Pandit A B. A review of imperative technologies for wastewater treatment I: oxidation technologies at ambient conditions. Advances in Environmental Research, 2004, 8 (3): 501-551.

[34] Yang Y, Wyatt I I, Travis D, Bahorsky M. Decolorization of dyes using UV/H_2O_2 photochemical oxidation. Textile Chemist and Colorist, 1998, 30 (4): 27-35.

[35] Peralta-Zamora P, Kunz A, de Moraes S G, Pelegrini R, de Campos Moleiro P, Reyes J, Duran N. Degradation of reactive dyes I. A comparative study of ozonation, enzymatic and photochemical processes. Chemosphere, 1999, 38 (4): 835-852.

[36] Slokar Y M, Majcen Le Marechal A. Methods of decoloration of textile wastewaters. Dyes and Pigments, 1998, 37 (4): 335-356.

[37] Robinson T, McMullan G, Marchant R, Nigam P. Remediation of dyes in textile effluent: a critical review on current treatment technologies with a proposed alternative. Bioresource technology, 2001, 77 (3): 247-255.

[38] Xu X R, Li H B, Wang W H, Gu J D. Decolorization of dyes and textile wastewater by potassium permanganate. Chemosphere, 2005, 59 (6): 893-898.

[39] Namboodri C G, Perkins W S, Walsh W K. Decolorizing dyes with chlorine and ozone: Part Ⅱ. American Dyestuff Reporter, 1994, 83 (4): 17-27.

[40] Kirby N, Singh D, Smyth F, Nigam P S N. Microbial decolourisation of textile dyes present in textile industries effluent//First International Symposium Biotechnology. Firence Italy, 1998.

[41] Hassan M M, Hawkyard C J. Decolorisation of effluent with ozone and re-use of spent dyebath//Environmental Aspects of Textile Dyeing. Elsevier, 2007: 149-190.

[42] Park T J, Lee K H, Jung E J, Kim C W. Removal of refractory organics and color in pigment wastewater with Fenton oxidation. Water Science and Technology, 1999, 39 (10-11): 189-192.

[43] Chu W, Ma C W. Quantitative prediction of direct and indirect dye ozonation kinetics. Water Research, 2000, 34 (12): 3153-3160.

[44] Aksu Z. Application of biosorption for the removal of organic pollutants: a review. Process Biochemistry, 2005, 40 (3): 997-1026.

[45] Beekeepers Y. Arising from reactive dyes in textile industry color fenton process remedy with. Istanbul: ITU Institute of Science, 2000.

[46] Mauskan J M. Advanced methods for treatment of textile industry effluents. Central Pollution Control Board Ministry of Environment of Forests, New Delhi, 2007.

[47] Sewekow U. Treatment of reactive dye effluents with hydrogen peroxide/iron (Ⅱ) sulphate. Melliand Textilberichte International Textile Reports, 1993, 74: E68-E68.

[48] Dutta K, Bhattacharjee S, Chaudhuri B, Mukhopadhyay S. Oxidative degradation of malachite green by Fenton generated hydroxyl radicals in aqueous acidic media. Journal of environmental science and health. Part A, Toxic/hazardous substances and environmental engineering, 2003, 38 (7): 1311-1326.

[49] Cuiping B, Wensheng X, Dexin F, Mo X, Dong G, Zhongxue G, Yanshui Z. Efficient decolorization of Malachite Green in the Fenton reaction catalyzed by [Fe (III) -salen] Cl complex. Chemical Engineering Journal, 2013, 215: 227-234.

[50] Vandevivere P C, Bianchi R, Verstraete W. Treatment and reuse of wastewater from the textile wet-processing industry: review of emerging technologies. Journal of Chemical Technology and Biotechnology: International Research in Process, Environmental And Clean Technology, 1998, 72 (4): 289-302.

[51] Banat I M, Nigam P, Singh D, Marchant R. Microbial decolorization of textile-dyecontaining effluents: A review. Bioresource Technology, 1996, 58 (3): 217-227.

[52] Saraf S, Vaidya V K. Comparative study of biosorption of textile dyes using fungal biosorbents. International Journal of Current Microbiology and Applied Science, 2015, 2: 357-365.

[53] Khan B, Goyal D (Guide). Microbial decolorization of triphenylmethane dyes. New York: Springer, 2007.

[54] Ali H. Biodegradation of synthetic dyes-a review. Water Air and Soil Pollution, 2010, 213 (1-4): 251-273.

[55] Dos Santos A B, Cervantes F J, van Lier J B. Review paper on current technologies for decolourisation of textile wastewaters: perspectives for anaerobic biotechnology. Bioresource technology, 2007, 98 (12): 2369-2385.

[56] Sharma N, Tiwari D P, Singh S K. The efficiency appraisal for removal of malachite green by potato peel and neem bark: isotherm and kinetic studies. International Journal, 2014, 5 (2): 83-88.

[57] Jasińska A, Rozalska S, Bernat P, Paraszkiewicz K, Dlugonski J. Malachite green decolorization by non-basidiomycete filamentous fungi of *Penicillium pinophilum* and *Myrothecium roridum*. International Biodeterioration and Biodegradation, 2012, 73: 33-40.

[58] Jasińska A, Bernat P, Paraszkiewicz K. Malachite green removal from aqueous solution using the system rapeseed press cake and fungus *Myrothecium roridum*. Desalination and Water Treatment, 2013, 51 (40-42): 7663-7671.

[59] Kaushik P, Malik A. Fungal dye decolourization: recent advances and future potential. Environment International, 2009, 35 (1): 127-141.

[60] Azad F N, Ghaedi M, Dashtian K, Hajati S, Goudarzi A, Jamshidi M. Enhanced simultaneous removal of malachite green and safranin O by ZnO nanorod-loaded activated carbon: modeling, optimization and adsorption isotherms. New Journal of Chemistry, 2015, 39 (10): 7998-8005.

[61] Banu S U N, Maheswaran G. Synthesis and characterization of nanocrystalline embedded activated carbon from Eichornia crassipes and its use in the removal of Malachite green. International Journal of Chem Tech Research, 2015, 8 (5): 158-169.

[62] Geetha P, Latha MS, Koshy M. Biosorption of malachite green dye from aqueous solution by calcium alginate nanoparticles: equilibrium study. Journal of Molecular Liquids, 2015, 212: 723-730.

[63] Sayğili H, Güzel F. Performance of new mesoporous carbon sorbent prepared from grape industrial processing wastes for malachite green and congo red removal. Chemical Engineering Research and Design, 2015, 100: 27-38.

[64] Nekouei F, Kargarzadeh H, Nekouei S, Tyagi I, Agarwal S, Kumar Gupta V. Preparation of Nickel hydroxide nanoplates modified activated carbon for Malachite Green removal from solutions: Kinetic, thermodynamic, isotherm and antibacterial studies. Process Safety and Environmental Protection, 2016, 102: 85-97.

[65] Song Y, Ding S, Chen S, Xu H, Mei Y, Ren J. Removal of malachite green in aqueous solution by adsorption on sawdust. Korean Journal of Chemical Engineering, 2015, 32 (12): 2443-2448.

[66] Mary A, Priya S, Devi N, Kousalya G N. Removal of malachite green from aqueous solution by adsorption studies using mosambi peal as an adsorbent. 2015, 7: 88-93.

[67] Velmurugan P, Rathinakumar V, Dhinakaran G. Dye removal from aqueous solution using low cost adsorbent. International journal of environmental sciences, 2011, 1 (7): 1492-1503.

[68] Mafra M R, Igarashi-Mafra L, Zuim DR, Vasques É, Ferreira MA. Adsorption of remazol brilliant blue on an orange peel adsorbent. Brazilian Journal of Chemical Engineering, 2013, 30 (3): 657-665.

[69] Singh D K, Rastogi K. Adsorptive removal of basic dyes from aqueous phase onto activated carbon of used tea leaves: a kinetic and thermodynamic study. Journal of environmental science and engineering, 2004, 46 (4): 293-302.

[70] Nagda G K, Ghole V S. Utilization of lignocellulosic waste from bidi industry for removal of dye from aqueous solution. International Journal of Environmental Research, 2008, 2 (4): 385-390.

[71] Chen H, Zhang L, Cheng X, Cheng S, Yan D. Adsorption behavior of malachite green from aqueous solution onto bamboo leaves biomass. Asian Journal of Chemistry, 2014, 26: 6579-6582.

[72] Kushwaha A K, Gupta N, Chattopadhyaya M C. Removal of cationic methylene blue and malachite green dyes from aqueous solution by waste materials of Daucus carota. Journal of Saudi Chemical Society, 2014, 18 (3): 200-207.

[73] Mullick A, Neogi S. Synthesis of potential biosorbent from used stevia leaves and its application for malachite green removal from aqueous solution: kinetics, isotherm and regeneration studies. RSC Advances, 2016, 6 (70): 65960-65975.

[74] Chowdhury S, Chakraborty S, Saha P. Biosorption of Basic Green 4 from aqueous solution by *Ananas comosus* (pineapple) leaf powder. Colloids and Surfaces B-Biointerfaces, 2011, 84 (2): 520-527.

[75] Arunadevi K, Venkatachalam R. A study on the preparation of adsorbent from Ananas comosus leaves plant by thermo chemical activation and its adsorption behavior of the removal of malachite green. Der Chemica Sinica, 2016, 7 (2): 14-22.

[76] Das P, Das P, Datta S. Continuous biosorption of Malachite Green by Ananus comosus (pineapple) leaf powder in a fixed bed reactor: experimental, breakthrough time and mathematical modeling. Desalination and Water Treatment, 2016, 57 (53): 25842-25847.

[77] Ho Y S, Chiu W T, Wang C C. Regression analysis for the sorption isotherms of basic dyes on sugarcane dust. Bioresource technology, 2005, 96 (11): 1285-1291.

[78] Patra T K, Sheeba K N, Augusta P, Jaisankar S. Experimental study on the adsorption of malachite green in simulated and real effluent by bio-based adsorbents. International Journal of Green Energy, 2015, 12 (11): 1189-1195.

[79] Hameed B H, El-Khaiary MI. Batch removal of malachite green from aqueous solutions by adsorption on oil palm trunk fibre: Equilibrium isotherms and kinetic studies. Journal of Hazardous Materials, 2008, 154 (1): 237-244.

[80] Jothirani R, Kumar PS, Saravanan A, Narayan AS, Dutta A. Ultrasonic modified corn pith for the sequestration of dye from aqueous solution. Journal of Industrial and Engineering Chemistry, 2016, 39: 162-175.

[81] Ren H, Zhang R, Wang Q, Pan H, Wang Y. Garlic root biomass as novel biosorbents for malachite green removal: Parameter optimization, process kinetics and toxicity test. Chemical Research in Chinese Universities, 2016, 32 (4): 647-654.

[82] Nethaji S, Sivasamy A, Thennarasu G, Saravanan S. Adsorption of Malachite Green dye onto activated carbon derived from *Borassus aethiopum* flower biomass. Journal of Hazardous Materials, 2010, 181 (1-3): 271-280.

[83] Chukki J, Shanthakumar S. Optimization of malachite green dye removal by *Chrysanthemum indicum* using response surface methodology. Environmental Progress and Sustainable Energy, 2016, 35 (5): 1415-1419.

[84] Santhi T, Manonmani S, Ravi S. Uptake of cationic dyes from aqueous solution by biosorption onto granular muntingia calabura. E-Journal of Chemistry, 2009, 6 (3): 737-742.

[85] Rahman IA, Saad B, Shaidan S, Sya Rizal ES. Adsorption characteristics of malachite green on activated carbon derived from rice husks produced by chemical-thermal process. Bioresource technology, 2005, 96 (14): 1578-1583.

[86] Chowdhury S, Mishra R, Saha P, Kushwaha P. Adsorption thermodynamics, kinetics and isosteric heat of adsorption of malachite green onto chemically modified rice husk. Desalination, 2011, 265 (1): 159-168.

[87] Chowdhury S, Saha PD. Scale-up of a dye adsorption process using chemically modified rice husk: optimization using response surface methodology. Desalination and Water Treatment, 2012, 37 (1-3): 331-336.

[88] Khan T A, Rahman R, Ali I, Khan E A, Mukhlif A A. Removal of malachite green from aqueous solution using waste pea shells as low-cost adsorbent-adsorption isotherms and dynamics. Toxicological and Environmental Chemistry, 2014, 96 (4): 569-578.

[89] Khattri S D, Singh M K. Removal of malachite green from dye wastewater using neem sawdust by adsorption. Journal of Hazardous Materials, 2009, 167 (1): 1089-1094.

[90] Garg V K, Kumar R, Gupta R. Removal of malachite green dye from aqueous solution by adsorption using agro-industry waste: a case study of *Prosopis cineraria*. Dyes and Pigments, 2004, 62 (1): 1-10.

[91] Papinutti L, Mouso N, Forchiassin F. Removal and degradation of the fungicide dye malachite green from aqueous solution using the system wheat bran-Fomes sclerodermeus. Enzyme and Microbial Technology, 2006, 39 (4): 848-853.

[92] Chowdhury S, Saha P. Sea shell powder as a new adsorbent to remove Basic Green 4 (Malachite Green) from aqueous solutions: Equilibrium, kinetic and thermodynamic studies. Chemical Engineering Journal, 2010, 164 (1): 168-177.

[93] Shamel A, Khoshraftar Z, Alayi R. Adsorption of cationic dye from aqueous solution onto sea shell as a adsorbent low-cost: kinetic studies. Der Pharma Chemica, 2016, 8: 60-66.

[94] Chowdhury S, Das (Saha) P. Mechanistic, Kinetic, and Thermodynamic Evaluation of Adsorption of Hazardous Malachite Green onto Conch Shell Powder. Separation Science and Technology, 2011, 46 (12): 1966-1976.

[95] Mittal A. Adsorption kinetics of removal of a toxic dye, Malachite Green, from wastewater by using hen feathers. Journal of Hazardous Materials, 2006, 133 (1): 196-202.

[96] Chowdhury S, Das P. Utilization of a domestic waste—eggshells for removal of hazardous malachite green from aqueous solutions. Environmental Progress and Sustainable Energy, 2012, 31 (3): 415-425.

[97] Chen K C, Wu J Y, Liou D J, Hwang S C J. Decolorization of the textile dyes by newly isolated bacterial strains. Journal of Biotechnology, 2003, 101 (1): 57-68.

[98] Przystas W, Zablocka-Godlewska E, Grabinska-Sota E. Biological removal of azo and triphenylmethane dyes and toxicity of process by-products. Water Air and Soil Pollution, 2012, 223 (4):

1581-1592.

[99] Singh A, Manju, Rani S, Bishnoi N R. Malachite green dye decolorization on immobilized dead yeast cells employing sequential design of experiments. Ecological Engineering, 2012, 47: 291-296.

[100] Safarik I, Ptackova L, Safarikova M. Adsorption of dyes on magnetically labeled baker's yeast cells. European Cells and Materials, 2002, 3 (2): 52-55.

[101] Ting A S Y, Lee M V J, Chow Y Y, Cheong S L. Novel exploration of *Endophytic Diaporthe* sp. for the biosorption and biodegradation of triphenylmethane dyes. Water Air and Soil Pollution, 2016, 227 (4): 109 (1-8).

[102] Chen S H, Cheow Y L, Ng S L, Ting A S Y. Removal of triphenylmethane dyes in single-dye and dye-metal mixtures by live and dead cells of metal-tolerant *Penicillium simplicissimum*. Separation Science and Technology, 2019, 55 (13): 2410-2420.

[103] Tsai W T, Chen H R. Removal of malachite green from aqueous solution using low-cost chlorella-based biomass. Journal of Hazardous Materials, 2010, 175 (1-3): 844-849.

[104] Sarayu K, Sandhya S. Current technologies for biological treatment of textile wastewater—a review. Applied biochemistry and biotechnology, 2012, 167 (3): 645-661.

[105] Gomaa O M, Selim N S, Linz J E. A possible role of aspergillus niger mitochondrial cytochrome c in Malachite Green reduction under calcium chloride stress. Cell Biochemistry and Biophysics, 2013, 67 (3): 1291-1299.

[106] Yan K, Wang H, Zhang X. Biodegradation of crystal violet by low molecular mass fraction secreted by fungus. Journal of Bioscience and Bioengineering, 2009, 108 (5): 421-424.

[107] Wang J, Qiao M, Wei K, Ding J, Liu Z, Zhang K Q, Huang X. Decolorizing activity of Malachite Green and its mechanisms involved in dye biodegradation by achromobacter xylosoxidans MG1. Journal of Molecular Microbiology and Biotechnology, 2011, 20 (4): 220-227.

[108] Karimi A, Aghbolaghy M, Khataee A, Bargh S S. Use of enzymatic bio-Fenton as a new approach in decolorization of malachite green. Scientific World Journal, 2012: 691569.

[109] Moldes D, Fernandez-Fernandez M, Angeles Sanroman M. Role of laccase and low molecular weight metabolites from trametes versicolor in dye decolorization. Scientific World Journal, 2012: 398725.

[110] Ren S, Guo J, Zeng G, Sun G. Decolorization of triphenylmethane, azo, and anthraquinone dyes by a newly isolated *Aeromonas hydrophila* strain. Applied Microbiology and Biotechnology, 2006, 72 (6): 1316-1321.

[111] Casas N, Parella T, Vicent T, Caminal G, Sarra M. Metabolites from the biodegradation of triphenylmethane dyes by Trametes versicolor or laccase. Chemosphere, 2009, 75 (10): 1344-1349.

[112] El-Sheekh M M, Gharieb M M, Abou-El-Souod G W. Biodegradation of dyes by some green algae and cyanobacteria. International Biodeterioration and Biodegradation, 2009, 63 (6): 699-704.

[113] Jing W, Jung B G, Kim K S, Lee Y C, Sung N C. Isolation and characterization of *Pseudomonas otitidis* WL-13 and its capacity to decolorize triphenylmethane dyes. Journal of Environmental Sci-

ences, 2009, 21 (7): 960-964.

[114] Kumar C G, Mongolia P, Joseph J, Sarma V U M. Decolorization and biodegradation of triphenylmethane dye, brilliant green, by *Aspergillus* sp isolated from Ladakh, India. Process Biochemistry, 2012, 47 (9): 1388-1394.

[115] Gomaa O M, Linz J E, Reddy C A. Decolorization of Victoria blue by the white rot fungus, *Phanerochaete chrysosporium*. World Journal of Microbiology and Biotechnology, 2008, 24 (10): 2349-2356.

[116] Saparrat M C, Mocchiutti P, Liggieri C S, Aulicino M B, Caffini N O, Balatti P A, Martínez M J. Ligninolytic enzyme ability and potential biotechnology applications of the white-rot fungus *Grammothele subargentea* LPSC no. 436 strain. Process Biochemistry, 2008, 43 (4): 368-375.

[117] Kalyani D C, Patil P S, Jadhav J P, Govindwar S P. Biodegradation of reactive textile dye Red BLI by an isolated bacterium *Pseudomonas* sp. SUK1. Bioresource Technology, 2008, 99 (11): 4635-4641.

[118] Parshetti G, Kalme S, Saratale G, Govindwar S. Biodegradation of malachite green by *Kocuria rosea* MTCC 1532. Acta Chimica Slovenica, 2006, 53 (4): 492-498.

[119] Shedbalkar U, Dhanve R, Jadhav J. Biodegradation of triphenylmethane dye cotton blue by *Penicillium ochrochloron* MTCC 517. Journal of Hazardous Materials, 2008, 157 (2-3): 472-479.

[120] Jang M S, Lee Y M, Kim C H, Lee J H, Kang D W, Kim S J, Lee Y C. Triphenylmethane reductase from *Citrobacter* sp. strain KCTC 18061P: purification, characterization, gene cloning, and overexpression of a functional protein in Escherichia coli. Applied and environmental microbiology, 2005, 71 (12): 7955-7960.

[121] Šafaříková M, Ptáčková L, Kibrikova I, Šafařík I. Biosorption of water-soluble dyes on magnetically modified *Saccharomyces cerevisiae* subsp. *uvarum* cells. Chemosphere, 2005, 59 (6): 831-835.

[122] Eichlerová I, Homolka L, Nerud F. Synthetic dye decolorization capacity of white rot fungus Dichomitus squalens. Bioresource Technology, 2006, 97 (16): 2153-2159.

[123] Chen C C, Liao H J, Cheng C Y, Yen C Y, Chung Y C. Biodegradation of crystal violet by *Pseudomonas putida*. Biotechnology Letters, 2007, 29 (3): 391-396.

[124] Chen C H, Chang C F, Ho C H, Tsai T L, Liu S M. Biodegradation of crystal violet by a *Shewanella* sp NTOU1. Chemosphere, 2008, 72 (11): 1712-1720.

[125] Safarik I, Rego L F T, Borovska M, Mosiniewicz-Szablewska E, Weyda F, Safarikova M. New magnetically responsive yeast-based biosorbent for the efficient removal of water-soluble dyes. Enzyme and Microbial Technology, 2007, 40 (6): 1551-1556.

[126] Abedin R M. Decolorization and biodegradation of crystal violet and malachite green by *Fusarium solani* (Martius) Saccardo. A comparative study on biosorption of dyes by the dead fungal biomass. Am Euras J Bot, 2008, 12: 17-31.

[127] Youssef A S, El-Sherif M F, El-Assar S A. Studies on the decolorization of malachite green by the local isolate *Acremonium kiliense*. Biotechnology, 2008, 7 (2): 213-223.

[128] Moturi B, Charya M S. Decolourisation of crystal violet and malachite green by fungi. Science World Journal, 2009, 4 (4): 28-33.

[129] Ayed L, Chaieb K, Cheref A, Bakhrouf A. Biodegradation and decolorization of triphenylmethane dyes by *Staphylococcus epidermidis*. Desalination, 2010, 260 (1-3): 137-146.

[130] Ghasemzadeh R, Kargar A, Lotfi M. Decolorization of synthetic textile dyes by immobilized white-rot fungus//International Conference on Chemical, Ecology and Environmental Sciences, Pattaya. 2011: 434-438

[131] Pandey R R, Pandey B V, Upadhyay R S. Characterization and identification of *Pseudomonas fluorescens* NCIM 2100 degraded metabolic products of Crystal Violet. African Journal of Biotechnology, 2011, 10 (4): 624-631.

[132] Parshetti G K, Parshetti S G, Telke A A, Kalyani D C, Doong R A, Govindwar S P. Biodegradation of crystal violet by *Agrobacterium radiobacter*. Journal of Environmental Sciences, 2011, 23 (8): 1384-1393.

[133] Torres J M O, Cardenas C V, Moron L S, Guzman A P A, dela Cruz T E E. Dye decolorization activities of marine-derived fungi isolated from Manila Bay and Calatagan Bay, Philippines. Philippine Journal of Science, 2011, 140 (2): 133-143.

[134] Yang X, Wang J, Zhao X, Wang Q, Xue R. Increasing manganese peroxidase production and biodecolorization of triphenylmethane dyes by novel fungal consortium. Bioresource Technology, 2011, 102 (22): 10535-10541.

[135] 杨秀清，李树仁，沈翀，王婧人．耐过氧化氢的锰过氧化物酶对三苯甲烷类染料的脱色（英文）．微生物学通报，2013，40 (8)：1356-1364.

[136] Cheriaa J, Khaireddine M, Rouabhia M, Bakhrouf A. Removal of triphenylmethane dyes by bacterial consortium. The Scientific World Journal, 2012 (4): 512454.

[137] Levin L, Papinutti L, Forchiassin F. Evaluation of Argentinean white rot fungi for their ability to produce lignin-modifying enzymes and decolorize industrial dyes. Bioresource technology, 2004, 94 (2): 169-176.

[138] Oranusi N A, Mbah A N. Utilisation of azo and triphenylmethane dyes as sole source of carbon, energy and nitrogen by *Bacillus* sp. African Journal of Applied Zoology and Environmental Biology, 2005, 7 (1): 87-94.

[139] Daneshvar N, Ayazloo M, Khataee A R, Pourhassan M. Biological decolorization of dye solution containing Malachite Green by microalgae *Cosmarium* sp. Bioresource Technology, 2007, 98 (6): 1176-1182.

[140] Deng D, Guo J, Zeng G, Sun G. Decolorization of anthraquinone, triphenylmethane and azo dyes by a new isolated *Bacillus cereus* strain DC11. International Biodeterioration & Biodegradation, 2008, 62 (3): 263-269.

[141] Du L N, Wang S, Li G, Wang B, Jia X M, Zhao Y H, Chen Y L. Biodegradation of malachite green by *Pseudomonas* sp. strain DY1 under aerobic condition: characteristics, degradation

products, enzyme analysis and phytotoxicity. Ecotoxicology, 2011, 20 (2): 438-446.

[142] Shedbalkar U, Jadhav J P. Detoxification of Malachite Green and Textile industrial effluent by *Penicillium ochrochloron*. Biotechnology and Bioprocess Engineering, 2011, 16 (1): 196-204.

[143] Vasdev K. Decolorization of triphenylmethane dyes by six white-rot fungi isolated from nature. Journal of Bioremediation and Biodegradation, 2011, 2 (5).

[144] Lv G Y, Cheng J H, Chen X Y, Zhang Z F, Fan L F. Biological decolorization of malachite green by *Deinococcus radiodurans* R1. Bioresource Technology, 2013, 144: 275-280.

[145] Mukherjee T, Das M. Degradation of malachite green by *Enterobacter asburiae* strain XJUHX-4TM. Clean Soil Air Water, 2014, 42 (6): 849-856.

[146] Sani R K, Banerjee U C. Decolorization of triphenylmethane dyes and textile and dye-stuff effluent by *Kurthia* sp. Enzyme and Microbial Technology, 1999, 24 (7): 433-437.

[147] Ayed L, Khelifi E, Ben Jannet H, Miladi H, Cheref A, Achour S, Bakhrouf A. Response surface methodology for decolorization of azo dye Methyl Orange by bacterial consortium Produced enzymes and metabolites characterization. Chemical Engineering Journal, 2010, 165 (1): 200-208.

[148] Ogugbue C J, Sawidis T, Oranusi N A. Bioremoval of chemically different synthetic dyes by Aeromonas hydrophila in simulated wastewater containing dyeing auxiliaries. Annals of Microbiology, 2012, 62 (3): 1141-1153.

[149] Khan S S, Arunarani A, Chandran P. Biodegradation of basic violet 3 and acid blue 93 by pseudomonas putida. Clean Soil Air Water, 2015, 43 (1): 67-72.

[150] Mishra S, Maiti A. The efficacy of bacterial species to decolourise reactive azo, anthroquinone and triphenylmethane dyes from wastewater: a review. Environmental Science and Pollution Research, 2018, 25 (9): 8286-8314.

[151] Chen C Y, Kuo J T, Cheng C Y, Huang Y T, Ho I H, Chung Y C. Biological decolorization of dye solution containing malachite green by Pandoraea pulmonicola YC32 using a batch and continuous system. Journal of Hazardous Materials, 2009, 172 (2-3): 1439-1445.

[152] Wu Y, Xiao X, Xu C, Cao D, Du D. Decolorization and detoxification of a sulfonated triphenylmethane dye aniline blue by Shewanella oneidensis MR-1 under anaerobic conditions. Applied Microbiology and Biotechnology, 2013, 97 (16): 7439-7446.

[153] Chen C C, Chen C Y, Cheng C Y, Teng P Y, Chung Y C. Decolorization characteristics and mechanism of Victoria Blue R removal by *Acinetobacter calcoaceticus* YC210. Journal of Hazardous Materials, 2011, 196: 166-172.

[154] Saratale R G, Saratale G D, Chang J S, Govindwar S P. Bacterial decolorization and degradation of azo dyes: a review. Journal of the Taiwan Institute of Chemical Engineers, 2011, 42 (1): 138-157.

[155] Guerra-Lopez D, Daniels L, Rawat M. *Mycobacterium* smegmatis mc (2) 155 fbiC and MSMEG_2392 are involved in triphenylmethane dye decolorization and coenzyme F-420 biosynthesis. Microbi-

ology-Sgm，2007，153：2724-2732.

[156] Wu J，Li L，Du H，Jiang L，Zhang Q，Wei Z，Wang X，Xiao L，Yang L. Biodegradation of leuco derivatives of triphenylmethane dyes by Sphingomonas sp CM9. Biodegradation，2011，22 (5)：897-904.

[157] Mukherjee T，Bhandari M，Halder S，Das M. Decolorization of malachite green by a novel strain of Enterobacter. Journal of the Indian Chemical Society，2012，89 (10)：1369-1374.

[158] Pan T，Ren S，Guo J，Xu M，Sun G. Biosorption and biotransformation of crystal violet by *Aeromonas hydrophila* DN322p. Frontiers of Environmental Science & Engineering，2013，7 (2)：185-190.

[159] Haritash A K，Kaushik C P. Biodegradation aspects of polycyclic aromatic hydrocarbons (PAHs)：A review. Journal of Hazardous Materials，2009，169 (1-3)：1-15.

[160] Kwasniewska K. Biodegradation of crystal violet (hexamethyl-p-rosaniline chloride) by oxidative red yeasts. Bulletin of environmental contamination and toxicology，1985，34 (3)：323-330.

[161] Moe N K T，Wilaipun P，Yonezuka K，Ishida W，Yano H，Terahara T，Imada C，Kamio M，Kobayashi T. Isolation and characterization of malachite green-removing yeast from a traditional fermented fishery product. Fisheries Science，2015，81 (5)：937-945.

[162] Yatome C，Ogawa T，Matsui M. Degradation of crystal violet by *Bacillus subtilis*. Journal of Environmental Science & Health Part A，1991，26 (1)：75-87.

[163] Yatome C，Yamada S，Ogawa T，Matsui M. Degradation of crystal violet by *Nocardia corallina*. Applied Microbiology and Biotechnology，1993，38 (4)：565-569.

[164] Rajeshkannan R，Rajasimman M，Rajamohan N. Removal of malachite green from aqueous solution using hydrilla verticillata-optimization，equilibrium and kinetic studies. International Journal of Civil and Environmental Engineering，2010，2 (4)：222-229.

[165] Vasanth Kumar K，Ramamurthi V，Sivanesan S. Biosorption of malachite green，a cationic dye onto *Pithophora* sp.，a fresh water algae. Dyes and Pigments，2006，69 (1)：102-107.

[166] Jerold M，Sivasubramanian V. Biosorption of malachite green from aqueous solution using brown marine macro algae *Sargassum swartzii*. Desalination and Water Treatment，2016，57 (52)：25288-25300.

[167] Han S，Han W，Chen J，Sun Y，Dai M，Zhao G. Bioremediation of malachite green by cyanobacterium *Synechococcus elongatus* PCC 7942 engineered with a triphenylmethane reductase gene. Applied Microbiology and Biotechnology，2020，104 (7)：3193-3204.

[168] Sharma D K，Saini H S，Singh M，Chimni S S，Chadha B S. Isolation and characterization of microorganisms capable of decolorizing various triphenylmethane dyes. Journal of basic microbiology，2004，44 (1)：59-65.

[169] Idaka E，Ogawa T，Yatome C，Horitsu H. Behavior of activated sludge with dyes. Bulletin of environmental contamination and toxicology，1985，35 (6)：729-734.

[170] Ogawa T，Idaka E，Yatome C. Studies on the treatment of the waste water containing dyestuffs by microor-

ganisms. Microbiology for Environment Cleaning. Published under the support of the Ministry of Education，grant，1978（212204-1977）：426-437.

[171] Ogawa T，Idaka E，Yatome C. Acclimation of activated sludge to dye. Bulletin of environmental contamination and toxicology，1981，26（1）：31-37.

[172] Ogawa T，Shibata M，Yatome C，Idaka E. Growth inhibition of *Bacillus subtilis* by basic dyes. Bulletin of environmental contamination and toxicology，1988，40（4）：545-552.

[173] Jadhav J P，Govindwar S P. Biotransformation of malachite green by *Saccharomyces cerevisiae* MTCC 463. Yeast，2006，23（4）：315-323.

[174] Deivasigamani C，Das N. Biodegradation of Basic Violet 3 by *Candida krusei* isolated from textile wastewater. Biodegradation，2011，22（6）：1169-1180.

[175] Nozaki K，Beh CH，Mizuno M，Isobe T，Shiroishi M，Kanda T，Amano Y. Screening and investigation of dye decolorization activities of basidiomycetes. Journal of bioscience and bioengineering，2008，105（1）：69-72.

[176] Sarnthima R，Khammuang S，Svasti J. Extracellular ligninolytic enzymes by *Lentinus polychrous* Lév. under solid-state fermentation of potential agro-industrial wastes and their effectiveness in decolorization of synthetic dyes. Biotechnology and Bioprocess Engineering，2009，14（4）：513-522.

[177] Jadhav S B，Yedurkar S M，Phugare S S，Jadhav J P. Biodegradation studies on Acid Violet 19，a triphenylmethane dye，by *Pseudomonas aeruginosa* BCH. Clean-Soil Air Water，2012，40（5）：551-558.

[178] Yan J，Niu J，Chen D，Chen Y，Irbis C. Screening of Trametes strains for efficient decolorization of malachite green at high temperatures and ionic concentrations. International Biodeterioration and Biodegradation，2014，87：109-115.

[179] Zhang C，Diao H，Lu F，Bie X，Wang Y，Lu Z. Degradation of triphenylmethane dyes using a temperature and pH stable spore laccase from a novel strain of *Bacillus vallismortis*. Bioresource Technology，2012，126：80-86.

[180] Moosvi S，Kher X，Madamwar D. Isolation，characterization and decolorization of textile dyes by a mixed bacterial consortium JW-2. Dyes and Pigments，2007，74（3）：723-729.

[181] Kandelbauer A，Guebitz G M. Bioremediation for the decolorization of textile dyes—a review//Lichtfouse E，Schwarzbauer J，Robert D. Environmental Chemistry：Green Chemistry and Pollutants in Ecosystems. Berlin：Springer，2005：269-288.

[182] Kunamneni A，Ballesteros A，Plou F J，Alcalde M. Fungal laccase—a versatile enzyme for biotechnological applications. Communicating current research and educational topics and trends in applied microbiology，2007，1：233-245.

[183] Champagne P P，Ramsay JA. Contribution of manganese peroxidase and laccase to dye decoloration by *Trametes versicolor*. Applied Microbiology and Biotechnology，2005，69（3）：276.

[184] Shin K S，Kim Y H，Lim J S. Purification and characterization of manganese peroxidase of the white-rot fungus *Irpex lacteus*. The Journal of Microbiology，2005，43（6）：503-509.

[185] Husain Q. Potential applications of the oxidoreductive enzymes in the decolorization and detoxification of textile and other synthetic dyes from polluted water: a review. Critical reviews in biotechnology, 2006, 26 (4): 201-221.

[186] Alam M Z, Mansor M F, Jalal K C A. Optimization of decolorization of methylene blue by lignin peroxidase enzyme produced from sewage sludge with *Phanerocheate chrysosporium*. Journal of Hazardous Materials, 2009, 162 (2-3): 708-715.

[187] dos Santos Bazanella G C, de Souza D F, Castoldi R, Oliveira R F, Bracht A, Peralta R M. Production of laccase and manganese peroxidase by *Pleurotus pulmonarius* in solid-state cultures and application in dye decolorization. Folia microbiologica, 2013, 58 (6): 641-647.

[188] Eichlerova I, Homolka L, Nerud F. Evaluation of synthetic dye decolorization capacity in Ischnoderma resinosum. Journal of Industrial Microbiology and Biotechnology, 2006, 33 (9): 759-766.

[189] Jiang M, Ten Z, Ding S. Decolorization of synthetic dyes by crude and purified laccases from coprinus comatus grown under different cultures: the role of major isoenzyme in dyes decolorization. Applied Biochemistry and Biotechnology, 2013, 169 (2): 660-672.

[190] Boer C G, Obici L, de Souza C G M, Peralta R M. Decolorization of synthetic dyes by solid state cultures of *Lentinula* (*Lentinus*) *edodes* producing manganese peroxidase as the main ligninolytic enzyme. Bioresource Technology, 2004, 94 (2): 107-112.

[191] Balan K, Sathishkumar P, Palvannan T. Decolorization of malachite green by laccase: Optimization by response surface methodology. Journal of the Taiwan Institute of Chemical Engineers, 2012, 43 (5): 776-782.

[192] Zhuo R, He F, Zhang X, Yang Y. Characterization of a yeast recombinant laccase rLAC-EN3-1 and its application in decolorizing synthetic dye with the coexistence of metal ions and organic solvents. Biochemical Engineering Journal, 2015, 93: 63-72.

[193] Jones J J, Falkinham J O. Decolorization of malachite green and crystal violet by waterborne pathogenic mycobacteria. Antimicrobial agents and chemotherapy, 2003, 47 (7): 2323-2326.

[194] Cha C J, Doerge D R, Cerniglia C E. Biotransformation of malachite green by the fungus Cunninghamella elegans. Applied and environmental microbiology, 2001, 67 (9): 4358-4360.

[195] Moturi B, Charya M A S. Strain improvement in dye decolourising mutants of Mucor mucedo by protoplast fusion. African Journal of Biotechnology, 2009, 8 (24): 6908-6912.

[196] Kalyani D C, Telke A A, Surwase S N, Jadhav S B, Lee J K, Jadhav J P. Effectual decolorization and detoxification of triphenylmethane dye malachite green (MG) by Pseudomonas aeruginosa NCIM 2074 and its enzyme system. Clean Technologies and Environmental Policy, 2012, 14 (5): 989-1001.

[197] Wang J, Gao F, Liu Z, Qiao M, Niu X, Zhang K Q, Huang X. Pathway and molecular mechanisms for malachite green biodegradation in exiguobacterium sp MG2. Plos One, 2012, 7 (12): e51808.

[198] Kim M H, Kim Y, Park H J, Lee J S, Kwak S N, Jung W H, Lee S G, Kim D, Lee Y C, Oh

T K. Structural insight into bioremediation of triphenylmethane dyes by *Citrobacter* sp. triphenylmethane reductase. Journal of Biological Chemistry，2008，283 (46)：31981-31990.

[199] Lu L，Zhao M，Liang S C，Zhao L Y，Li D B，Zhang B B. Production and synthetic dyes decolourization capacity of a recombinant laccase from Pichia pastoris. Journal of Applied Microbiology，2009，107 (4)：1149-1156.

[200] Chengalroyen M D，Dabbs E R. Identification of a gene responsible for amido black decolorization isolated from Amycolatopsis orientalis. World Journal of Microbiology and Biotechnology，2013，29 (4)：625-633.

[201] Fu X Y，Zhao W，Xiong A S，Tian Y S，Zhu B，Peng R H，Yao Q H. Phytoremediation of triphenylmethane dyes by overexpressing a Citrobacter sp triphenylmethane reductase in transgenic Arabidopsis. Applied Microbiology and Biotechnology，2013，97 (4)：1799-1806.

第 3 章 嗜水气单胞菌对三苯基甲烷染料的脱色

3.1 脱色菌分离鉴定与脱色酶表达

任随周等从印染废水的厌氧污泥中筛选到一株广谱染料脱色细菌，经鉴定后命名为嗜水气单胞菌 DN322[1]。该菌可同时将偶氮和三苯基甲烷染料脱色[2]。据报道，有几种白腐真菌具有降解三苯基甲烷染料的能力，这些真菌在生物反应器中经常用于偶氮染料的脱色和降解[3]。虽然真菌对废水的处理通常非常耗时[4]，但是固定化酶可以在不需要添加生长底物的情况下对废水进行脱色和循环利用。白腐真菌对结晶紫的降解依赖于木质素降解体系[5]。其他真菌，包括 birds' nest fungi，*Pycnoporus sanguineus*，*Trametes versicolor*，*Dichomitus squalens* 和 *Phlebia* sp. 等，均能通过漆酶降解结晶紫[3,6-8]。为了使这一过程在经济上更有吸引力，需要显著降低这些酶的成本[9]。

当利用枯草芽孢杆菌 IF0 13719 和珊瑚色诺卡氏菌 iam112121 细胞降解结晶紫时，鉴定出一种产物米氏酮（MK）[10,11]。其他细菌，包括 *Kurthia* sp.，*Citrobacter* sp.，*Mycobacteria avium* strain A5，*Alcaligenes* sp.，*Aeromonas* sp.，*Pseudomonas* sp. 和肠道细菌也被报道有对三苯基甲烷染料脱色的能力。任随周等从嗜水气单胞菌 DN322 中分离纯化出一种脱色酶，并鉴定为一种含血红素的加氧酶，对几种三苯基甲烷染料的脱色和降解具有催化作用[12,13]。

3.1.1 菌种分离与鉴定

(1) 形态与特征

DN322 菌株为革兰氏阴性菌，杆状，可运动，单极性鞭毛，大小为（0.7～1.0)μm×(1.0～3.5)μm（图 3-1)[1]。在 LB 平板上生长的菌落为浅黄色至粉红色，表面光滑，边缘整齐，菌落直径为 0.5～3mm。

(2) 生理生化鉴定

DN322 可在 4～40℃、pH 值为 6.0～10.0 下生长，最适温度和 pH 值分别为 25～30℃ 和 7.0。该菌兼性厌氧，氧化酶和过氧化氢酶呈阳性，Voges-

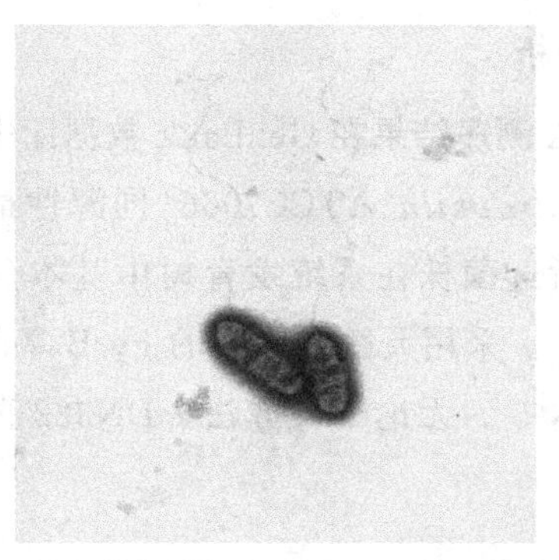

图 3-1　嗜水气单胞菌 DN322 的电子显微镜照片（15000×）[1]

Proskauer 检测为阳性。该菌株可发酵葡萄糖产气，发酵半胱氨酸产 H_2S。API 20NE 系统鉴定为种级，鉴定结果与模式菌嗜水气单胞菌 ATCC7966^{T}进行了比较。DN322 的生化和生理特性与嗜水气单胞菌 ATCC7966^{T}一致，详见表 3-1[1]。

表 3-1　DN322 与模式菌的生理生化鉴定结果[1]

项目名称	项目缩写	DN322	模式菌
尿素酶	URE	－	－
色氨酸	TRP	＋	＋
D-对硝基苯基-β-D-吡喃半乳糖苷的水解	ONPG	＋	＋
葡萄糖:酸化	GLU	＋	＋
L-阿拉伯糖	ARA	＋	＋
精氨酸二水解酶	ADH	＋	＋
甘露醇	MAN	＋	＋
七叶苷:水解	ESC	＋	＋
乙酸苯酯	PAC	＋	＋
己二酸	ADI	＋	＋
癸酸盐	CAP	＋	＋
明胶	GEL	＋	＋
细胞色素氧化酶	OX	＋	＋
葡萄糖:同化	GLU	＋	＋
葡萄糖酸盐	GNT	＋	＋
苹果酸	MLT	＋	＋
麦芽糖	MAL	＋	＋
甘露糖	MNE	＋	＋
N-乙酰氨基葡萄糖	NAG	＋	＋
柠檬酸盐	CIT	＋	＋
KNO_3	NO_3	＋	＋

注：“＋”为阳性，“－”为阴性。

(3) 系统发育学分析

DN322 的 16S rRNA 测序结果在 GenBank 数据库中利用 BlAST 与已登记基因序列比对后，与 *A.hydrophila* ATCC7966^{T}同源性最高，达到 99%。系统发育树如图 3-2 所示[1]。所选菌株在系统发育树中基本分为 5 个类群，DN322 菌株在嗜水气单胞菌类群中。采用灵敏度更高的 gyrB 基因序列与相关菌株建立的系统进化树如图 3-3 所示[1]。无论哪种方法，DN322 菌株都被分类到嗜水气单胞菌属中。

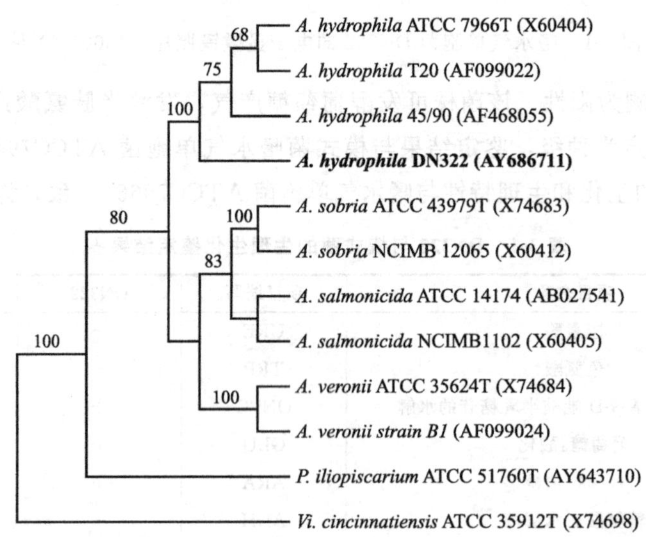

图 3-2 基于 16S rDNA 序列比对的系统进化树[1]

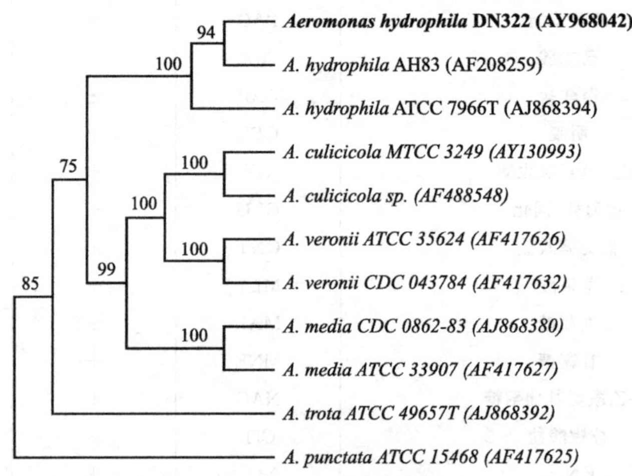

图 3-3 基于 gyrB 基因序列比对的系统进化树[1]

3.1.2 脱色酶的分离与纯化

(1) 产酶条件及酶的表达类型分析

DN322 在 LB 培养基中培养过后，分别提取细胞质和细胞膜组分用于结晶紫的脱色，结果发现前者的活性是后者的 10 倍以上，因此可以推测脱色酶位于细胞质中[1]。而投加 NADH 或 NADPH（25mmol/L）后，染料脱色效率大幅提高，说明该脱色酶属于 NADH/NADPH 依赖型[1]。

在不含有染料的 LB 培养基中培养该菌株，每隔一段时间取样测脱色酶活性。结果发现脱色酶的表达类型是明显的生长偶联型，与细胞生长呈显著正相关[1]。16h 后细胞从对数生长期进入稳定期时，酶活性达到最大值。为确定该脱色酶是组成型表达还是诱导型表达，在 LB 培养基中加入结晶紫后测酶活性差异[1]。结果显示加与不加结晶紫对酶的表达和活性没有影响，这说明该脱色酶是组成型表达[1]。

(2) 脱色酶的分离与纯化

利用 LB 培养 DN322 细胞 16h 后，收集细胞并高压匀浆破碎，得到粗酶液[1]。然后用 Q-Sepharose 离子交换柱层析，收集每个蛋白峰洗脱液。具有酶活性的蛋白峰经超滤离心浓缩后，利用 HiTrap Blue 亲和层析柱专一性纯化脱色酶。FPLC 洗脱液脱色活性测定及活性显示染色结果表明，嗜水气单胞菌 DN322 细胞裂解液中只有一种脱色酶蛋白质具有活性。变性 SDS-PAGE 凝胶（图 3-4）和 IEF 凝胶中存在一条带，证实了这一结果。通过非变性 PAGE 测定酶的分子

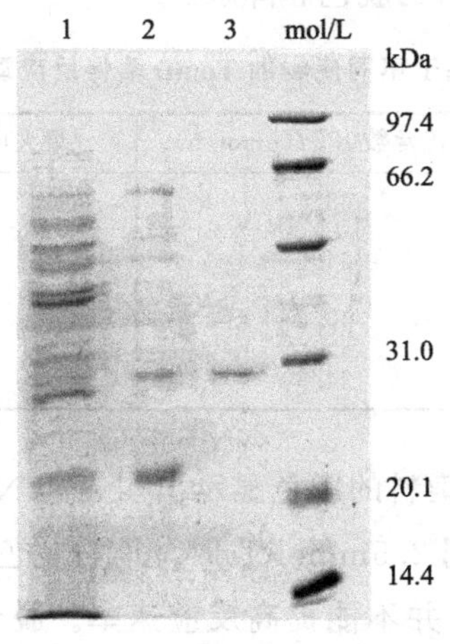

图 3-4 纯化脱色酶的 SDS-PAGE 电泳图[1]

质量约为87500Da。SDS-PAGE估计每个亚基的分子质量约为29400Da。这些结果表明这种酶是三聚体[13]。用等电聚焦法测定脱色酶的等电点为pH=5.6。经上述三步纯化之后，脱色酶活性提高了12.7倍，得率超过50%，命名为三苯胺染料脱色酶（triphenylamine dyes decolorization enzyme，TpmD)[1]。

(3) 酶学性质

脱色酶在41～59℃范围内表现出较高的活性，最佳反应温度为51℃。当温度高于62℃时，酶活性大大降低。在4℃冷藏3个月，剩余酶活性仍在80%以上。酶活性在30℃以下相对稳定，35℃保存7天后酶活性损失50%。当温度超过55℃时，脱色酶活性性迅速下降，在65℃下放置3h，酶活性损失超过70%。以剩余酶活性为50%的温度作为标准，TpmD的半失活温度为62℃[1,13]。

TpmD的最佳脱色pH值为7.5。pH值处于5.5～9.0范围内时，相对酶活性超过50%。在pH值在5.0以下或9.5以上时，酶活性残留极低。脱色酶在pH值在6.0～9.0范围内时相对稳定，残留酶活性保持在85%以上。

(4) 底物特异性

10min之内，TpmD将四种三苯基甲烷染料，孔雀石绿、灿烂绿、碱性品红及结晶紫分别脱色了87%、90%、95%和100%[1,13]。脱色酶对四种具有相似结构的染料脱色率相似，说明该酶具有底物特异性。进一步采用Lineweaver-Burk作图分析TpmD的动力学参数，结果如表3-2显示。米氏常数K_m最小的结晶紫为TpmD的最适底物[1,13]。然而，与嗜水气单胞菌DN322的完整细胞不同，纯化的酶不能催化偶氮染料的脱色。这说明嗜水气单胞菌DN322中存在另一个酶系统，负责偶氮染料的脱色和降解。

表3-2 基于不同底物的TpmD酶促反应动力学参数

三苯基甲烷染料	米氏常数K_m/(μmol/L)	最大反应速率V_{max}/[μmol/(L·s)]
孔雀石绿	68.5	115.6
灿烂绿	54.2	82.8
碱性品红	40.6	74.1
结晶紫	24.3	19.6

TpmD对三苯基甲烷染料的脱色显示出明显的NADH或NADPH依赖性。当NADH浓度从0增加到2.5mmol/L时，染料脱色率线性增加[1]。在该浓度以上，进一步添加NADH并不能提高反应速率。脱色酶可以同时使用NADH和NADPH作为共基质，NADH对酶的活性提高更为有效。NADH和NADPH

的 K_m 分别为 511μmol/L 和 792μmol/L，蛋白质的 V_{max} 分别为 6870μmol/(L·s) 和 5020μmol/(L·s)。在一定的 NADH 浓度下，脱色反应速率取决于酶的用量。而黄素辅酶 FAD 或 FMN 对酶活性无影响。另外，纯化酶的脱色活性还取决于分子氧的存在，这一点通过对所有被测三苯基甲烷染料脱色过程中溶解氧浓度的监测得到证实。脱色过程中溶解氧浓度迅速下降，说明脱色反应消耗了氧分子。当反应液用氮气吹脱后，未观察到脱色现象。这些结果表明，分子氧是纯化酶脱色三苯甲烷染料所必须的，该酶是一种 NADH 依赖的加氧酶[1]。

实验检测了 14 种金属离子对 TpmD 酶活性的影响，发现 Hg^{2+}、Fe^{3+}、Cu^{2+}、Ag^{+}、Fe^{2+} 和 Al^{3+} 可抑制脱色，Zn^{2+} 有部分抑制作用，并未发现刺激酶活性的金属离子[1]。反应体系中分别添加 2～10mmol/L 的电子传递抑制剂鱼藤酮（Rotenone）、抗霉素 A（Antimycin A）或 NaN_3 后，TpmD 酶活性几乎不受影响。自由基抑制剂维生素 C 和苯甲酸钠对脱色酶活性影响显著。浓度达到 15mmol/L 的维生素 C 和 200mmol/L 的苯甲酸钠几乎完全抑制了酶的活性，说明 TpmD 的脱色过程中可能产生了自由基。另外，细胞色素 P450 的三种竞争性抑制剂（喹啉、吡啶和甲吡酮）都对 TpmD 的脱色产生了明显的抑制作用，显示该脱色酶可能与细胞色素 P450 有关[1]。

（5）结晶紫脱色产物分析

用分光光度法和气相色谱-质谱联用（GC-MS）技术对嗜水气单胞菌 DN322 的脱色酶降解结晶紫进行了研究，发现结晶紫水溶液在 590nm、550nm、303nm 和 249nm 处有四个峰（图 3-5）。在反应混合物中加入 NADH 后，590nm、550nm 和 303nm 处的吸收峰强度逐渐降低，340nm 和 259nm 处出现两个新的吸

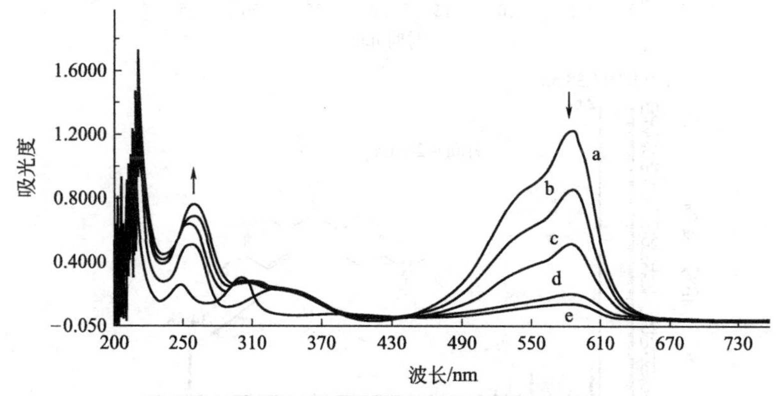

图 3-5 结晶紫脱色过程的光谱变化[1]

a 为磷酸盐缓冲溶液中的结晶紫；b 为向 a 中加入 NADH 后的结晶紫；c～e 分别为向 b 中加入脱色酶反应 2min、4min 和 6min 后的结晶紫。从 a 到 e，右边的峰逐渐降低，左边的峰逐渐升高

收峰，分别为 NADH 和 NAD 的最大吸收峰。脱色过程中，590nm 和 550nm 处的峰在 10min 后完全消失，340nm 处的峰（NADH 的最大峰）持续存在，259nm 处的峰（NAD 的最大峰）强度增大。

GC-MS 分析表明，氧化脱色反应的主要产物在气相色谱仪上的保留时间（34.88min）与标准品 MK 相同（图 3-6）。相应组分的峰出现在 m/z（相对丰

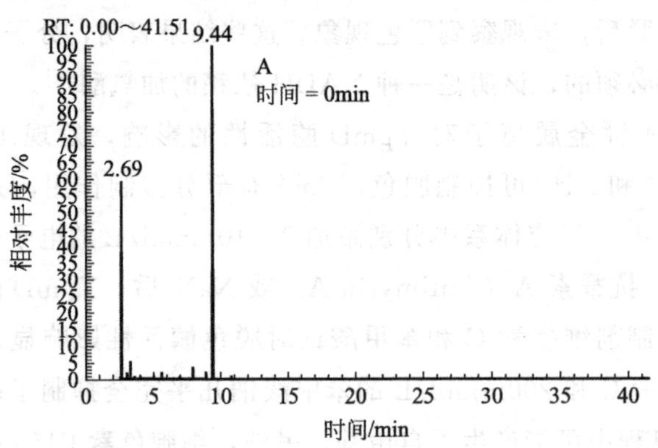

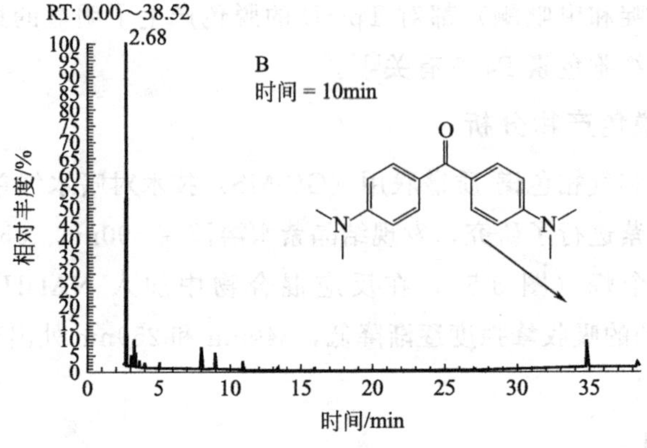

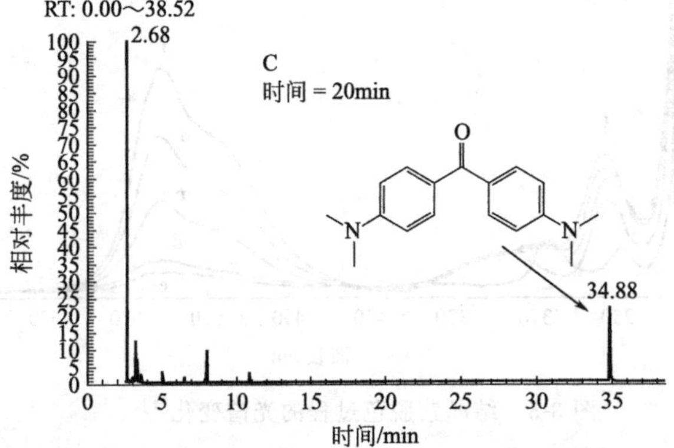

图 3-6 结晶紫脱色产物 MK 的 GC-MS 谱图[1]

度)：269(M+1,19)，268(M$^+$,100)，267(M−1,23)，251(18)，224(29)，148(72)。在脱色反应开始时未检测到MK，但在脱色过程中逐渐积累。结果表明，MK是纯化的加氧酶对结晶紫脱色反应的主要产物之一[1]。

(6) 碳源的影响

利用LB或者添加了不同碳源的M9培养基培养DN322细胞后，取等量细胞破碎提取粗酶液并检测其酶活[1]。如图3-7所示，LB或者M9培养基添加丙三醇后培养的DN322细胞酶活性明显更高，而添加了乳酸钠和丙酮酸钠M9的酶活性非常低。如前所述，TpmD脱色酶是组成型表达酶。碳源不同导致酶的活性不同，可能与该酶在细胞内的生理功能有关。

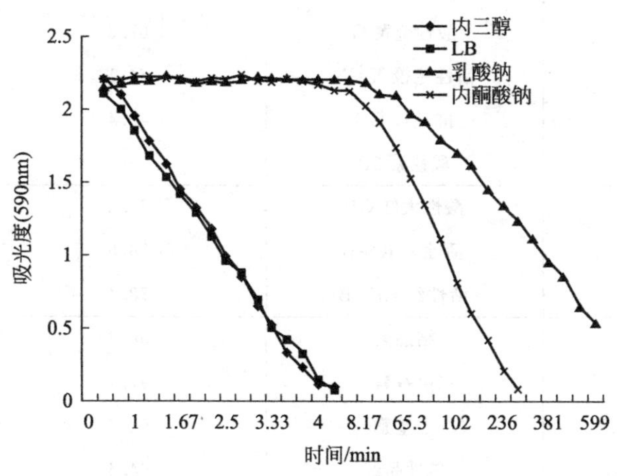

图3-7　碳源对TpmD酶活性的影响[1]

3.1.3 染料脱色特性

(1) DN322对不同种类染料的脱色能力

DN322是一种广谱脱色菌，具有对多种不同染料的脱色能力，如偶氮、蒽醌和三苯基甲烷染料等[1]。表3-3列出了DN322可以脱色的染料[1]。包括单偶氮和双偶氮在内的大部分偶氮染料都能被DN322脱色，且大部分的24h脱色率可以达到80%。酸性苋菜红的脱色率最高，6h能脱色96.4%。其次是活性艳红M-8B和酸性大红GR，14h可脱色95%以上。对酸性红GG的脱色效率较低，48h仅能达到40%左右。唯一没有脱色活性的染料是酸性紫2R，60h内无脱色效果。DN322对多种三苯基甲烷染料具有脱色活性和解毒能力。

50mg/L 的结晶紫、甲基紫、乙基紫、碱性品红、灿烂绿和孔雀石绿在 6～9h 内，都能达到 90%以上的脱色率。但对于弱酸艳蓝 FFB、酸性湖蓝 A 和酸性艳绿 B 三个分子量更大的三苯基甲烷染料，DN322 的脱色效率有限，这可能和染料分子的高级空间结构与 TpmD 酶催化活性中心的结合阻力有很大关系。另外，DN322 对部分蒽醌、二氮杂半菁阳离子和硫化染料也具有较好的脱色活性。

表 3-3 DN322 对不同种类染料的脱色能力[1]

染料种类	染料名称	最大脱色率/%	脱色时间/h
单偶氮	酸性苋莱红	96.4	6
	活性艳红 M-8B	95.5	14
	酸性大红 BS	88.4	12
	酸性金黄 G	67.3	24
	活性艳橙 X-GN	64.1	36
	酸性红 GG	40.4	48
	酸性紫 2R	—	60
双偶氮	酸性大红 GR	95.2	24
	活性黑 KN-B	86.5	24
	活性红 KE-3B	78.4	24
三苯基甲烷	结晶紫	98.3	6
	甲基紫	97.4	6.5
	乙基紫	92.6	8
	碱性品红	94.4	9
	灿烂绿	93.7	7
	孔雀石绿	95.2	8
	弱酸艳蓝 FFB	36.5	24
	酸性湖蓝 A	24.1	24
	酸性艳绿 B	17.9	24
蒽醌	活性艳兰 K-GR	85.3	36
	酸性蓝 25	60.4	52
	分散红 11	92.4	12
	酸性蓝 56	21	52
二氮杂半菁阳离子	阳离子兰 X-GRRL	95.7	5.2
硫化	硫化黑 9	91.2	4
	硫化亮绿	95.4	5
	硫化宝蓝	87.5	6

(2) 影响染料脱色的环境因素

由于偶氮还原酶一般受分子氧的抑制，因此多数微生物都需要厌氧或者兼性厌氧条件才能脱色偶氮染料。对于 DN322，严格厌氧条件下的偶氮脱色明显好于兼性厌氧，而好氧条件下几乎不脱色[1]。另一个直接证据是，偶氮染料的脱色从脱色管底部开始逐渐上移直至脱色完成，这说明厌氧更有利于脱色。而三苯基甲烷染料脱色的需氧情况完全不同[1]。好氧摇床的脱色速率最快，静置培养的兼性厌氧次之，严格厌氧条件几乎不脱色。以结晶紫为例，利用氮气将培养基中的氧气吹脱后再放入厌氧培养箱脱色，脱色率极低。这说明三苯基甲烷染料的 DN322 脱色是好氧型的，脱色酶活性对氧存在依赖作用。

分别利用偶氮染料的酸性苋菜红、酸性大红 GR 和活性艳红 M-8B 检测 DN322 的合适脱色温度，得知在 20～37℃之间，脱色效果较好[1]。四种三苯基甲烷染料（结晶紫、碱性品红、灿烂绿和孔雀石绿）的脱色结果与偶氮染料基本一致。

当 pH 值在 3.0～12.0 范围内，DN322 对偶氮染料和三苯基甲烷染料的脱色都表现出中性酸碱度需求[1]。脱色培养基的 pH 值低于 5.0 或高于 10.0 时，脱色率显著降低。并且，随着脱色反应的进行，培养基的 pH 值向中性迁移，说明中性 pH 值更有利于染料脱色和细胞生长。

染料浓度也会影响 DN322 的脱色。偶氮染料中的酸性苋菜红和酸性大红 GR 的浓度在 200mg/L 和 300mg/L 以上时，DN322 的脱色能力明显减弱[1]。而三苯基甲烷染料的结晶紫浓度达到 500mg/L 时才完全抑制 DN322 的脱色。其他三苯基甲烷染料如孔雀石绿和灿烂绿等浓度达 1000mg/L 时才对脱色产生影响。

接种量明显影响染料脱色的迟滞期。50mg/L 酸性苋菜红的实验表明，0.2%的接种量，DN322 需要接近 7h 才能进入脱色对数期[1]。而接种量达到 15%后，染料脱色可不经历迟滞期，立即开始。对于三苯基甲烷染料，接种量达到 10%以后，迟滞期也明显变短，脱色速率也显著提高。

3.1.4 脱色酶的克隆与表达

(1) 蛋白质重组表达系统

蛋白质的合成和调节取决于细胞的功能需求。蛋白质的设计图存储在 DNA 中，并通过高度调控的转录过程进行解码，以产生信使 RNA（mRNA）。然后由 mRNA 编码的信息翻译成蛋白质。转录是信息从 DNA 到 mRNA 的转移，翻译是基于 mRNA 指定序列的蛋白质合成。

在原核生物中，转录和翻译过程同时发生。mRNA 的翻译甚至在完全合成成熟的 mRNA 转录之前就已开始。基因的这种同时转录和翻译称为偶联转录和翻译。在真核生物中，这些过程在空间上是分开的，并按顺序发生，转录发生在细胞核中，翻译或蛋白质合成发生在细胞质中[14]。

转录在原核生物和真核生物中均发生三个步骤：起始、延伸和终止。当解开双链 DNA 以允许 RNA 聚合酶结合时，转录开始。一旦开始转录，RNA 聚合酶就会从 DNA 中释放出来。转录受激活剂和阻遏物以及真核生物中染色质结构的调节。在原核生物中，不需要对 mRNA 进行特殊修饰，并且甚至在转录完成之前就开始翻译。然而，在真核生物中，mRNA 会进一步加工（剪接）以去除内含子，在 5′端添加一个帽，在 3′端添加多个腺嘌呤，以生成 poly A 尾巴。然后将修饰的 mRNA 输出到细胞质中进行翻译。翻译后，以各种方式修饰多肽以完成其结构，指定其位置或调节其在细胞内的活性。翻译后修饰（PTM）是化学结构的各种添加或改变，是整个细胞生物学的关键特征[15]。

为了研究特定蛋白质的生物学特性，研究人员通常需要一种生产（制造）目的功能蛋白质的方法。考虑到蛋白质的大小和复杂性，通常利用活细胞作为载体，根据提供的遗传模板来构建和表达蛋白质。

使用成熟的重组 DNA 技术，DNA 易于合成或体外构建。因此，可以将具有或不具有附加报道基因或亲和标签序列的特定基因的 DNA 模板构建为蛋白质表达的模板。由这种 DNA 模板产生的蛋白质称为重组蛋白质。

重组蛋白质表达的传统策略包括用包含模板的 DNA 载体转染细胞，然后培养细胞以使其转录和翻译所需的蛋白。通常，需要裂解细胞以提取表达的蛋白质，用于随后的纯化。原核和真核蛋白质表达系统被广泛使用。系统的选择取决于蛋白质的类型、功能活性的要求和所需的产量。这些表达系统包括哺乳动物、昆虫、酵母、细菌、藻类和无细胞体系。每个系统都有其优点和挑战，重组蛋白的成功表达，选择适合的正确系统至关重要。

哺乳动物蛋白质表达：哺乳动物表达系统可用于产生具有天然结构和活性的哺乳动物蛋白质。哺乳动物表达系统是表达哺乳动物蛋白质的优选系统，可用于生产抗体、复杂蛋白质和用于基于功能细胞测定的蛋白质。但是，哺乳动物表达系统需要非常苛刻的养殖条件。哺乳动物表达系统可通过稳定的细胞系生产蛋白质，表达构建载体并整合到宿主基因组中。虽然稳定的细胞系可用于多个实验，但悬浮细胞培养可在一到两周内产生大量蛋白质。这些高产量的哺乳动物表达系统利用悬浮培养，可以产生 g/L 级别的产量。此外，与其他表达系统相比，这些蛋白质具有更多的天然折叠和翻译后修饰，例如糖基化。

昆虫蛋白质表达：可以将昆虫细胞用于与哺乳动物系统相似的高水平蛋白质表达。有几种重组杆状病毒系统，可将其用在昆虫细胞中表达目的蛋白质。这些系统可以轻松扩大规模，并适用于高密度悬浮培养，大规模表达与天然哺乳动物蛋白质功能相似的蛋白质。尽管产量很高，但重组杆状病毒的生产可能很耗时，而且培养条件比原核生物系统更苛刻。

细菌蛋白质表达：细菌蛋白质表达系统之所以受欢迎，是因为细菌易于培养，生长迅速并产生高产量的重组蛋白质。然而，在细菌中表达的真核蛋白质通常是无功能的，因为细胞不具备完成所需的翻译后修饰或分子折叠的能力。同样，许多蛋白质形成不溶性包涵体，如果没有变性剂和随后烦琐的蛋白质复性步骤，很难恢复活性。

植物表达系统：植物提供了廉价且低技术的重组蛋白质大规模表达的手段。来自各种植物（例如玉米、烟草、水稻、甘蔗甚至马铃薯的块茎）的许多细胞已用于蛋白质表达。植物系统与哺乳动物细胞表达系统具有许多相同的特征和加工要求，包括大多数复杂的翻译后修饰。然而，由于植物组织本身结构复杂，因此从植物中提取和纯化重组蛋白可能昂贵且费时。为了避免这些问题，科学家利用植物根部自然分泌的植物肽标记重组蛋白质，可以更轻松地获得和纯化所需蛋白质。尽管这是一项新兴技术，但植物细胞已被用于表达各种蛋白质，包括抗体和药物，特别是白介素。

酵母表达系统：酵母是产生大量重组真核蛋白质的良好表达系统。尽管许多酵母可以用于蛋白质表达，但是由于其在遗传学和生物化学中用作模型生物，所以酿酒酵母是最可靠和最常用的物种。当使用酿酒酵母时，研究人员经常将重组蛋白质置于半乳糖诱导型启动子（GAL）的控制之下。其他常用的启动子分别包括磷酸盐和铜诱导的 PHO5 和 CUP1 启动子。酵母细胞在确定的培养基中生长，可以轻松地进行发酵，从而实现大规模、稳定的蛋白质生产。通常，与哺乳动物细胞相比，酵母表达系统更易于操作且更便宜。并且与细菌系统不同，它通常能够修饰复杂的蛋白质。然而，酵母细胞的生长速度比细菌细胞慢，并且通常需要优化生长条件。

无细胞表达系统：在无细胞表达系统中，使用负责转录和翻译的核糖体、RNA 聚合酶、tRNA、核糖核苷酸和氨基酸，在体外组装蛋白质。无细胞表达系统是在一个反应中快速组装多种蛋白质的理想选择。这些系统的主要优点是它们能够组装具有标记或修饰的氨基酸的蛋白质，这些蛋白质可用于不同的下游应用中。然而，无细胞表达系统昂贵且技术复杂具有挑战性。

(2) 脱色酶在大肠杆菌中的表达

聚合酶链反应（PCR）扩增以 pET22-*tpmD* 为模板，p1 和 p2 作为上下游底物，琼脂糖凝胶电泳显示 900bp 处出现扩增产物，大小与 *tpmD* 基因接近[16]。将质粒在大肠杆菌 BL21（DE3）感受态细胞中融合表达，并用 AP 平板法筛选单菌落。将脱色酶表达良好的重组菌离心并收集细胞，超声波破碎后离心得到粗酶液，再经过亲和层析得到纯化蛋白质。采用碱性品红、结晶紫、孔雀石绿和灿烂绿检测脱色酶的活性，发现与 DN322 野生菌相比，重组菌的脱色酶粗提物活性更高[16]。

(3) 脱色酶在大肠杆菌中的无诱导表达

采用希瓦氏菌 S12 的 SND 启动子（NADPH 脱氢酶基因启动子序列）与 DN322 编码的 *tpmD* 基因融合构建无诱导的大肠杆菌基因工程菌[17]。以四种三苯基甲烷染料为脱色底物，研究了含有 SND 启动子的基因工程菌、DN322 野生菌和不含 SND 启动子的基因工程菌三者之间的全细胞脱色能力。结果发现，脱色效率依次降低。希瓦氏菌 S12 的 SND 启动子极大地提升了基因工程菌的染料脱色效率。并且，该重组菌在水产养殖废水孔雀石绿的脱色中有良好的工程应用潜力[17]。

(4) 乳糖替代 IPTG 诱导脱色酶在大肠杆菌中的表达

利用乳糖代替昂贵的乳糖操纵子诱导剂异丙基硫代-β-D-半乳糖苷（IPTG），有效降低工程菌构建规模化应用的成本[18]。结果表明，乳糖在细胞生长的任何阶段添加都有一定的诱导作用。其中，培养 3h 左右添加效果最明显。乳糖与 IPTG 相比，后者的诱导作用起效更快，而乳糖由于需要细胞膜转运体系辅助进入细胞而诱导相对滞后。同时，乳糖诱导工程菌表达的脱色酶在大肠杆菌细胞内并未形成包涵体，而是形成具有活性的可溶蛋白质。相对野生 DN322 菌种，重组菌的酶活性提高了 12 倍以上[18]。

(5) 毕赤酵母中脱色酶的表达及有机溶剂和抑制剂的影响

除细菌表达系统外，*tpmD* 基因也尝试了在酵母表达系统中的表达[19]。与大肠杆菌不同，毕赤酵母表达的脱色酶可直接分泌到细胞外，这对酶的高产和分离纯化极为有利。测试了四种有机溶剂对重组酶活性的影响。其中甲醇影响程度最低，二甲基亚砜次之，高浓度的丙酮和乙醇几乎完全抑制了酶活性。常用的酶抑制剂，如 EDTA、L-半胱氨酸和叠氮化钠，对酶活性影响很小，但 EDTA 浓度升高后影响增大。阴离子表面活性剂十二烷基硫酸钠（SDS）对酶活性的影响最大，低浓度下即可完全抑制酶活性。

(6) DTT 对重组酶脱色的影响

① DTT 对孔雀石绿的脱色[20]

250μmol/L 孔雀石绿只需要 62.5μmol/L 的二硫苏糖醇（DTT）就可以达到很好的脱色率（97%），DTT 的用量仅为染料的 25%，并且脱色反应在 1min 以内即可完成。即，平均每 1mol 的 DTT 可脱色 4mol 的孔雀石绿。

DTT 的还原力受 pH 值的影响，只有在一定 pH 值的情况下能够发挥还原作用。在溶液 pH 值小于 5 时，溶液 H^+ 浓度较高，DTT 活性基团—SH 无法脱去质子，还原力受到抑制。当溶液 pH 值大于 5 时，溶液 H^+ 浓度降低，—S^- 浓度增大，DTT 还原能力增强，脱色能力增强。此时，DTT 的孔雀石绿脱色能力维持在 80%以上。

脱色产物经高效液相色谱（HPLC）分析，未能检测到孔雀石绿的吸收峰。说明孔雀石绿的脱色效果比较彻底。通过 HPLC 分析发现脱色产物含有隐性孔雀石绿。这表明 DTT 对孔雀石绿的脱色机理是通过在孔雀石绿中心碳原子上加氢，从而将孔雀石绿还原为无色的隐性孔雀石绿。在 DTT 的孔雀石绿脱色反应过程中，通过肉眼观察可发现随着反应的进行，反应溶液中有白色絮状沉淀产生。初步怀疑此沉淀为 DTT 与隐性孔雀石绿形成的水不溶性的配合物。这将直接导致反应溶液中隐性孔雀石绿的减少，并进一步导致 HPLC 检测到的隐性孔雀石绿峰高的降低。白色沉淀的具体化学成分有待于后续实验的进一步证明。

根据以上实验结果，推测 DTT 脱色孔雀石绿的途径为：1mol DTT 通过还原反应可将 4mol 孔雀石绿还原为隐性孔雀石绿，生成的 4mol 隐性孔雀石绿进一步与 1mol 氧化态的 DTT 配合生成白色沉淀。

② DTT 对重组酶脱色的影响

张培培等利用 DTT 替代 NADH 作为 TpmD 重组酶的辅酶，考察其影响[19]。发现 DTT 作为辅酶，可达到 NADH 90%左右的效果。单据使用 TpmD 重组酶和 DTT 都无法使染料脱色。DTT 浓度对 TpmD 脱色也有影响。仅 0.5mmol/L 的 DTT 可将 TpmD 的脱色率达到 85%以上。当其浓度超过 2.5mmol/L，染料脱色率超过 90%。脱色过程中，溶氧浓度始终平稳，这与 NADH 辅助下的好氧脱色完全不同[19]。利用全波长扫描发现染料和 DTT 的吸收峰在脱色完成后几近消失，而并未出现新峰。说明这一脱色过程的机理与 NADH 作为辅酶时明显不同[19]。

3.2
菌种培养与产物分析

结晶紫是一类常见的三苯基甲烷染料，常用于印染、制药、食品和化妆品等工业[21]。结晶紫结构稳定并具有微生物及哺乳类细胞毒性[4]。由于结晶紫毒性较强，传统的废水处理系统难以有效处理。一些物理[22]和化学[23]的脱色方法已有报道。但这些方法存在高消耗、产生大量污泥和后续处理困难等不足。

运用强化的生物处理方法来治理这类染料污染是一个很好的选择。三苯基甲烷染料的脱色菌[8,24,25]和染料生物降解方法已有相关报道[4]。与传统的治理方法相比，生物脱色具有环境友好和消耗低等优点。Wu 等发现 *Pseudomonas otitidis* WL-13（*P. otitidis* WL-13）通过物理吸附对结晶紫进行脱色[26]，但 *Citrobacter* sp. KCTC 18061P 则将结晶紫转化为它的无色形式[27]。通常情况下，对于纯菌，其脱色方式为物理吸附或者生物转化[4]。

嗜水气单胞菌 DN322 是笔者团队分离出的一株脱色谱很广的染料脱色菌[2]。Ren 等发现它对三苯基甲烷染料、偶氮染料和蒽醌染料都具有很强的脱色能力[2]。在本实验中，我们从嗜水气单胞菌 DN322 的一个 5 年冻存管中重新分离出一株它的子代并命名为 DN322p。与 DN322 和其他染料降解菌不同的是，DN322p 可以同时通过生物吸附和生物转化的方法来脱色结晶紫。为描述 DN322p 细胞的吸附能力，我们用紫外-可见分光光度法（UV-Vis spectrophotometer）对二氯甲烷萃取液进行全波长扫描分析。除吸附外，DN322p 能将结晶紫转化为无色产物。脱色产物经薄层色谱法（thin-layer chromatography，TLC）和 HPLC 检测并确定为隐性结晶紫。同时，实验进一步比较了 DN322p 活菌体和死菌体的吸附能力。

3.2.1 菌种与培养基

通过菌液稀释涂平板挑选到了嗜水气单胞菌 DN322 的子代。收集 50 个单克隆抗体并纯化两次后比较各自的结晶紫脱色能力。脱色最快的菌命名为嗜水气单胞菌 DN322p 并用于进一步的实验。

DNA提取：使用细菌DNA提取试剂盒提取复筛菌种基因组DNA，OD260/280＝2.017。通过通用引物对16s rRNA进行克隆。PCR反应体系为(50μL)：0.4μL Taq酶，2μL缓冲液，0.3μL脱氧核糖核苷三磷酸（dNTP)，2μL模板DNA，0.5μL引物8f，0.5μL引物1094R，44.3μL水。PCR反应条件(25个循环)：95℃，7min；94℃，30s；48℃，30s；72℃，2min；22℃暂停。PCR产物经电泳检测后，将明亮无拖尾的反应产物拿给测序公司测序。得到的测序结果经软件拼接后，在NCBI上经BLACT进行比对分析。

细胞生长培养基为LB培养基，组成为10g/L蛋白胨，5g/L酵母提取物，10g/L NaCl。

脱色培养基包括12.5g/L NaCl，28.5g/L $Na_2HPO_4 \cdot 12H_2O$，13.3g/L KH_2PO_4，2.5g/L $(NH_4)_2SO_4$。

3.2.2 培养条件与脱色实验

在100mL的三角瓶中装40mL的LB培养基。接种DN322p后，在200r/min、30℃下振荡培养12h。然后，在8000r/min、20℃下离心5min收集细胞。

收集的细胞用于染料脱色实验。将脱色培养基中的细胞和结晶紫浓度分别调节到OD600 1.0和50mg/L。取100mL调节好的培养基置入250mL三角瓶中，在200r/min和30℃下震荡培养。对照体系不加细胞。

将121℃下灭菌30min的菌体用于死菌体吸附实验。其细胞浓度同样调节到OD600 1.0。

3.2.3 分析方法

每0.5h，取1mL脱色液在8000r/min下离心收集细胞及上清液。上清液在590nm下经用分光光度计测光密度值。脱色率由式(3-1) 计算：

$$\text{脱色率}=\left(1-\frac{\text{上清液中结晶紫浓度}}{\text{结晶紫初始浓度}}\right)\times 100\% \tag{3-1}$$

离心分离后，上清液和菌体分别用二氯甲烷萃取后，用UV-Vis进行全波长扫描分析吸收峰的变化。扫描范围为200～800nm。

脱色产物通过TLC进行初步分析。色谱条件是：丙醇：水：冰醋酸＝90：9：1（体积比）[27]。

脱色产物经HPLC进行最终定性和定量分析。HPLC色谱条件为：以90%

的甲醇水溶液作为流动相，流速为1.0mL/min。结晶紫和隐性结晶紫溶解在二氯甲烷中作为标样。

3.3 三苯基甲烷染料的脱色特性

3.3.1 聚合酶链反应鉴定复筛菌株

PCR产物的凝胶电泳检测，如图3-8所示。条带明亮，无拖尾。说明克隆序列比较单一，凝胶用于序列测定。

图3-8 PCR产物的凝胶电泳检测

测序公司返回的序列经拼接后，在NCBI上与DN322的16s rRNA进行BLAST比对，结果如下：

Score=1555 bits (842)，Expect=0.0，

Identities=869/880 (99%)，Gaps=10/880 (1%)，

Strand=Plus/Minus

```
Query  4     CACCGTGGT-AACGCCCTCCCG-AGGTTAAGCTATCTACTTCTGGTGCAACCCACTCCCA  61
             ||||||||| |||||||||||| |||||||||||||||||||||||||||||||||||||
Sbjct  1424  CACCGTGGTAAACGCCCTCCCGAAGGTTAAGCTATCTACTTCTGGTGCAACCCACTCCCA  1365

Query  62    TGGTGTGACGGGCGGTGTGTACAAGGCCCGGGAACGTATTCACCGCAACATTCTGATTTG  121
             ||||||||||||||||||||||||||||||||||||||||||||||||||||||||||||
Sbjct  1364  TGGTGTGACGGGCGGTGTGTACAAGGCCCGGGAACGTATTCACCGCAACATTCTGATTTG  1305

Query  122   CGATTACTAGCGATTCCGACTTCACGGAGTCGAGTTGCAGACTCCGATCCGGACTACGAC  181
             ||||||||||||||||||||||||||||||||||||||||||||||||||||||||||||
Sbjct  1304  CGATTACTAGCGATTCCGACTTCACGGAGTCGAGTTGCAGACTCCGATCCGGACTACGAC  1245

Query  182   GCGCTTTTTGGGATTCGCTCACTATCGCTAGCTTGCAGCCCTCTGTACGCGCCATTGTAG  241
             ||||||||||||||||||||||||||||||||||||||||||||||||||||||||||||
Sbjct  1244  GCGCTTTTTGGGATTCGCTCACTATCGCTAGCTTGCAGCCCTCTGTACGCGCCATTGTAG  1185

Query  242   CACGTGTGTAGCCCTGGCCGTAAGGGCCATGATGACTTGACGTCATCCCCACCTTCCTCC  301
             ||||||||||||||||||||||||||||||||||||||||||||||||||||||||||||
Sbjct  1184  CACGTGTGTAGCCCTGGCCGTAAGGGCCATGATGACTTGACGTCATCCCCACCTTCCTCC  1125

Query  302   GGTTTATCACCGGCAGTCTCCCTTGAGTTCCCACCATTACGTGCTGGCAACAAAGGACAG  361
             ||||||||||||||||||||||||||||||||||||||||||||||||||||||||||||
Sbjct  1124  GGTTTATCACCGGCAGTCTCCCTTGAGTTCCCACCATTACGTGCTGGCAACAAAGGACAG  1065

Query  362   GGGTTGCGCTCGTTGCGGGACTTAACCC-AACATCTCACGACACGAGCTGACGACAGCCA  420
             |||||||||||||||||||||||||||| |||||||||||||||||||||||||||||||
Sbjct  1064  GGGTTGCGCTCGTTGCGGGACTTAACCCAAACATCTCACGACACGAGCTGACGACAGCCA  1005

Query  421   TGCAGCACCTGTGTTCTGATTCCCGAAGGCACTCCCGTATCTCTACAGGATTCCAGACAT  480
             ||||||||||||||||||||||||||||||||||||||||||||||||||||||||||||
Sbjct  1004  TGCAGCACCTGTGTTCTGATTCCCGAAGGCACTCCCGTATCTCTACAGGATTCCAGACAT  945

Query  481   GTCAAGGCCAGGTAAGGTTCTTCGCGTTGCATCGAATTAAACCACATGCTCCACCGCTTG  540
             ||||||||||||||||||||||||||||||||||||||||||||||||||||||||||||
Sbjct  944   GTCAAGGCCAGGTAAGGTTCTTCGCGTTGCATCGAATTAAACCACATGCTCCACCGCTTG  885

Query  541   TGCGGGCCCCCGTCAATTCATTTGAGTTTTAACCTTGCGGCCGTACTCCCCAGGCGGTCG  600
             ||||||||||||||||||||||||||||||||||||||||||||||||||||||||||||
Sbjct  884   TGCGGGCCCCCGTCAATTCATTTGAGTTTTAACCTTGCGGCCGTACTCCCCAGGCGGTCG  825
```

```
Query  601   ATTTAACGCGTTAGCTCCGGAAGCCACGTCTCAAGGACACAGCCTCCAAATCGACATCGT  660
             ||||||||||||||||||||||||||||||||||||||||||||||||||||||||||||
Sbjct  824   ATTTAACGCGTTAGCTCCGGAAGCCACGTCTCAAGGACACAGCCTCCAAATCGACATCGT  765

Query  661   TTACGGCGTGGACTACCAGGGTATCTAATCCTGTTTGCTCCCCACGCTTTCGCACCTGAG  720
             ||||||||||||||||||||||||||||||||||||||||||||||||||||||||||||
Sbjct  764   TTACGGCGTGGACTACCAGGGTATCTAATCCTGTTTGCTCCCCACGCTTTCGCACCTGAG  705

Query  721   CGTCAGTCTTTTGTCCAGGGGGCCGCCTTCGCCACCGGTATTCCTCCAGATCTCTACGCA  780
             ||||||||||| ||||||||||||||||||||||||||||||||||||||||||||||||
Sbjct  704   CGTCAGTCTTT-GTCCAGGGGGCCGCCTTCGCCACCGGTATTCCTCCAGATCTCTACGCA  646

Query  781   TTTCACCCGCTACACCCTGGAATTCTAcccccccTCTACAAGACTCTAGCTGGACCAGTT  840
             ||||||| |||||||| |||||||||||||||| ||||||||||||||||||||| ||||
Sbjct  645   TTTCACC-GCTACACC-TGGAATTCTACCCCCC-TCTACAAGACTCTAGCTGGAC-AGTT  590

Query  841   TTAAATGCAATTTCCAGGTTTGAGCCCGGGGGCTTTCACA  880
             |||||||||||| ||||||| |||||||||| ||||||||
Sbjct  589   TTAAATGCAATTCCCAGGTT-GAGCCCGGGG-CTTTCACA  552
```

结果发现，DN322p 和 DN322 的 16s rRNA 相似度大于 99%。由此可以确定 DN322p 为 DN322 的子代。

3.3.2 细胞沉淀与上清液的颜色变化

如图 3-9 所示，仅 0.5h 后 50mg/L 结晶紫溶液的脱色率达到 80%左右。细

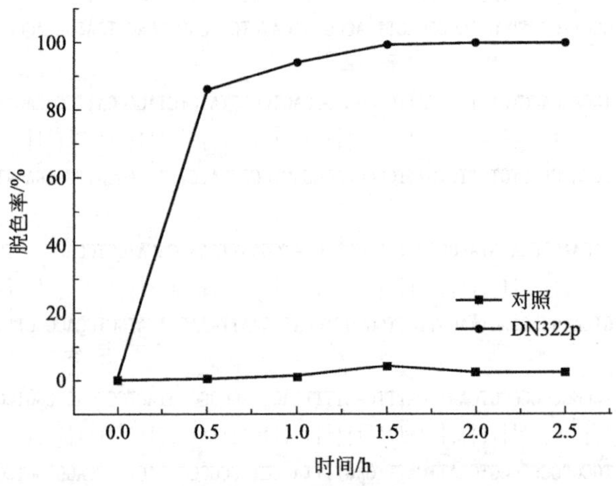

图 3-9 结晶紫的 DN322p 脱色曲线

胞沉淀被染色［图 3-10(a)］。2h 后，结晶紫的脱色率接近 100%。上清液随培养时间的延长而逐渐褪色。收获的细胞经二氯甲烷萃取后可清晰看到细胞被染色［图 3-10(b)］。

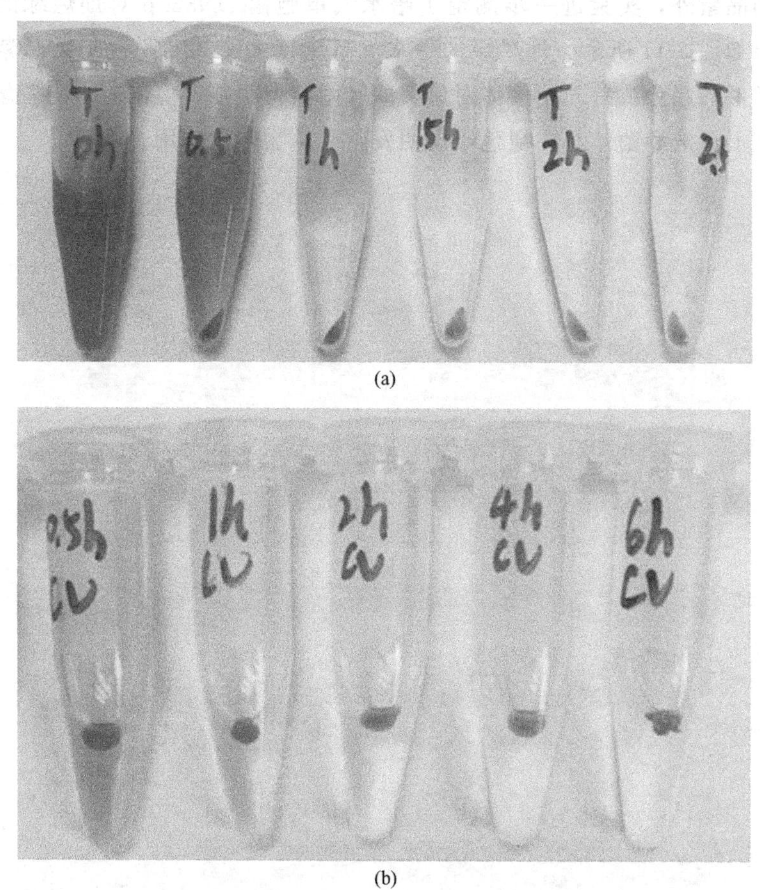

图 3-10　结晶紫的 DN322p 脱色过程

(a) 每隔 0.5h，取 1mL 溶液经 8000r/min 离心后拍照；(b) 细胞沉淀经 1mL 二氯甲烷萃取后拍照

通常情况下，细菌的染料脱色主要通过两种方式：吸附和/或生物降解。当染料通过吸附作用被脱色时，菌体会被染色并且不褪色；当染料通过生物降解方式脱色时，菌体保持无色。细菌 *P. otitidis* WL-13[26] 和 *Citrobacter* sp. KCTC 18061P[27] 的染料脱色机理刚好印证了此理论。但是 DN322p 脱色结晶紫表现出一种联合的脱色机理。

如图 3-9 所示，当染料脱色率为 80%时，菌体染色很深［图 3-10(a)］，这说明 DN322p 对结晶紫具有吸附能力。菌体的二氯甲烷萃取液中，菌体的颜色也十分鲜艳，很难被萃取到有机相，说明这种吸附作用十分牢固［图 3-10(b)］。然而，细胞的颜色并非一直不变。随着脱色时间的延长，菌体颜色由紫色逐渐变为灰色（图 3-10）。这说明菌体进一步将结晶紫转化成了其他化合物。

除结晶紫外，实验进一步测定了嗜水气单胞菌 DN322p 对灿烂绿的脱色情况。结果如图 3-11 所示，灿烂绿的嗜水气单胞菌 DN322p 脱色与结晶紫脱色呈现出相类似的趋势。首先，菌体对染料的吸附能力很强。但同时，随着培养时间的延长，上清液和菌体都在褪色，说明发生了生物转化作用。

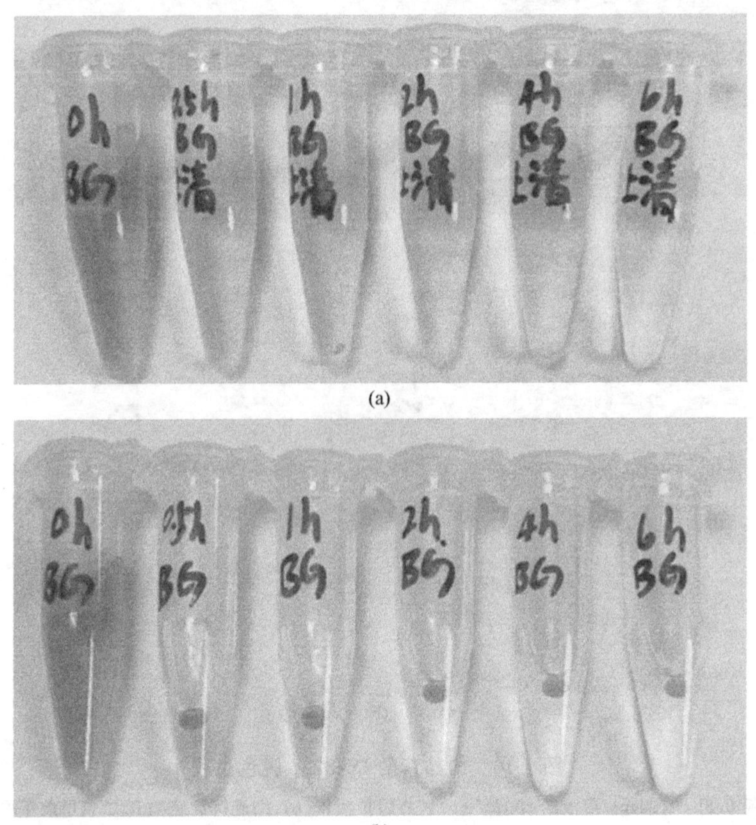

(a)

(b)

图 3-11　灿烂绿的 DN322p 脱色过程

(a) 每隔一段时间，取 1mL 溶液经 8000r/min 离心后拍照；(b) 细胞沉淀经 1mL 二氯甲烷萃取后拍照

3.3.3　紫外可见分光光度计分析染料脱色过程

UV-Vis 用于分析结晶紫及其脱色产物的吸收光谱变化。细胞沉淀及上清液的二氯甲烷萃取液的全波长吸收光谱如图 3-12 所示。上清液在 590nm 处的特征

吸收峰显著降低，并且在260nm处出现一个新峰。2h后，590nm处的吸收峰彻底消失，而260nm处的新峰达到最大值［图3-12(a)］。菌体的二氯甲烷萃取液表现出相似的趋势［图3-12(b)］。

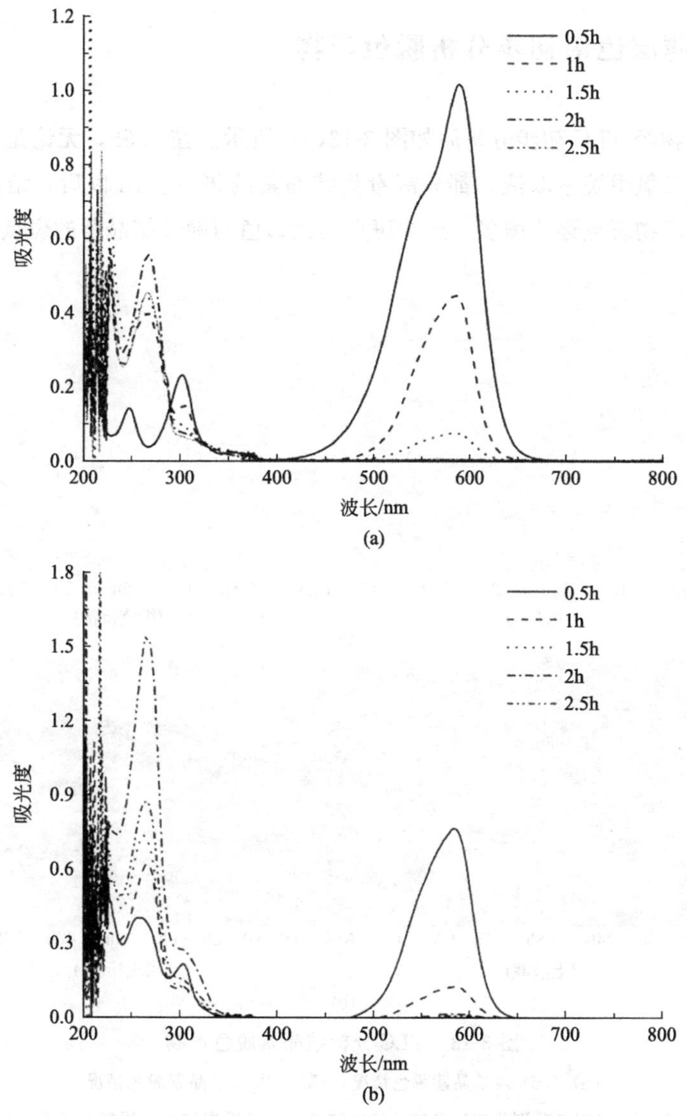

图3-12　结晶紫脱色过程中的紫外-可见吸收光谱

(a) 上清液的二氯甲烷萃取液；(b) 细胞沉淀的二氯甲烷萃取液

根据结晶紫在590nm处特征吸收峰的消失和260nm处出现新的特征吸收峰，UV-Vis分析的结果表明结晶紫被DN322p转化为其他化合物。此化合物在260nm处有特征吸收峰。Asad等阐明，当三苯基甲烷染料被脱色时，如果伴随着旧峰的消失和新峰的产生这种现象，那么脱色方式就是生物转化[28]。因此，通过以上两组实验可以有效说明，DN322p 同时通过生物吸附和生物转化的方式脱色结晶紫。

3.3.4 薄层色谱初步分析脱色产物

脱色产物经 TLC 初步分析后如图 3-13(a) 所示。在 0.5h，无论是上清液还是菌体沉淀的二氯甲烷萃取液，都只能看到结晶紫的斑点。1h 以后，结晶紫的斑点逐渐消失而产物斑点逐步增多。产物斑点的比移值与隐性结晶紫的斑点一致。

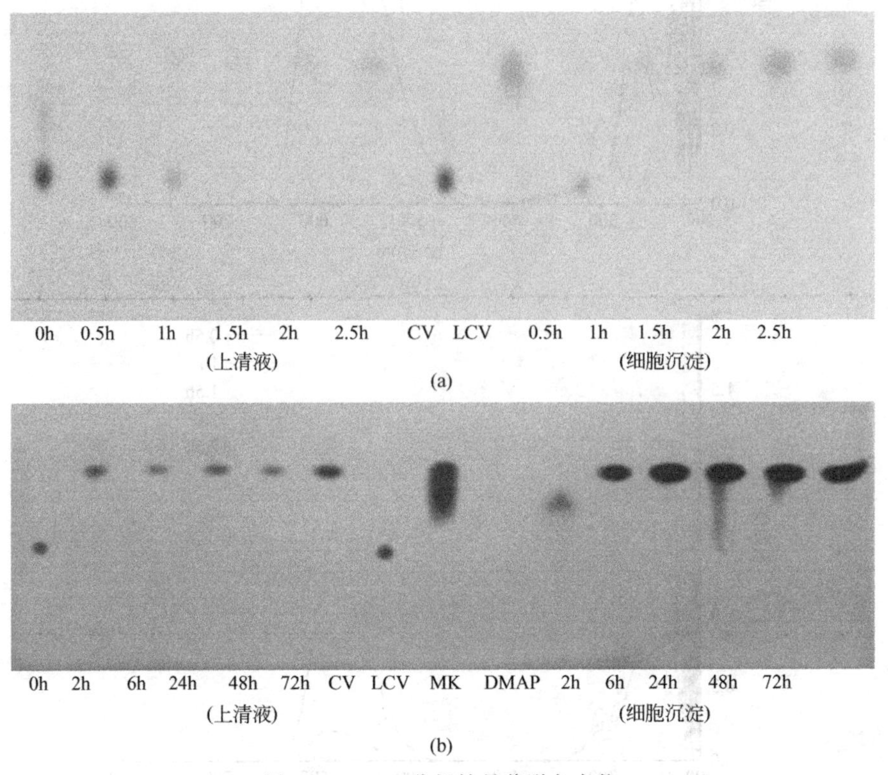

图 3-13 TLC 分析结晶紫脱色产物

(a) 2.5h 内结晶紫脱色情况；(b) 72h 内结晶紫脱色情况

CV、LCV、MK 和 DMAP 分别代表结晶紫、隐性结晶紫、米氏酮和 4-二甲氨基苯酚的标准样品对照

新产生的斑点与隐性结晶紫的比移值相同，说明结晶紫很有可能被 DN322p 转化为隐性结晶紫。

实验过程中，考虑到染料被进一步降解的可能性，将培养时间延长到 72h 后，再次通过 TLC 检测染料的残留与转化，结果如图 3-13(b) 所示。72h 以后，产生的隐性结晶紫含量几乎没有发生变化。说明菌体没能将隐性结晶紫进一步降解为其他产物。

TLC 检测［图 3-13(b)］中发现，在隐性结晶紫的上方也出现了不明显的斑点，与 MK 的位置相当。但是，通过 HPLC 和 GC-MS 分析无法检测到这种物质，也没有检测到与 MK 标样相同的物质。限于实验条件，该物质暂时未知，其确切的化学结构及性质还需实验进一步验证。

3.3.5 高效液相色谱分析脱色产物

通过 HPLC 分析可知，结晶紫和隐性结晶紫的保留时间分别为 10.086min 和 15.409min。2.5h 后，通过 UV-Vis 检测可知结晶紫基本脱色完全（图 3-12）。终产物的 HPLC 分析显示已无法检测到结晶紫的吸收峰［图 3-14(a)］。但是在 15.417min 处出现新峰，此峰与隐性结晶紫的保留时间相同，说明是同一物质［图 3-14(b)］。结果表明，隐性结晶紫可能是结晶紫的主要脱色产物。

3.3.6 质量平衡

图 3-15 计算了 50mg/L 结晶紫经 DN322p 脱色后结晶紫和隐性结晶紫的质量平衡。如图所示，脱色后，上清液中结晶紫含量不足 1%（质量分数），而产生的隐性结晶紫约占 7%（质量分数）。细胞沉淀用二氯甲烷萃取后分析 DN322p 对染料的吸附情况。大约 18%（质量分数）的隐性结晶紫和少于 0.5%（质量分数）的结晶紫被检测到。大于 70%（质量分数）的结晶紫在上清液和菌体的二氯甲烷萃取液中检测不到。据估计，这部分结晶紫或者其代谢产物（隐性结晶紫或其他产物）被吸附在菌体上，难以被二氯甲烷萃取出来。因此，这部分化合物的组成无法定论。

3.3.7 活菌体和死菌体的吸附能力对比

死菌体与活菌体调节到相同 OD600 之后，细胞经高压灭活，对比分析死菌体和活菌体对 50mg/L 结晶紫的吸附能力。结果如图 3-16 所示，死菌体呈现出与活菌体一样的吸附趋势。在 OD600 0.936 时，死菌体的脱色率为 83.5%，活菌体的脱色率达到 94.3%，活菌体略高。但是活菌体在测试其细胞的染料吸附能力时，无法避免它的生物转化能力。因此，活菌体和死菌体在染料吸附能力上，基本一致。

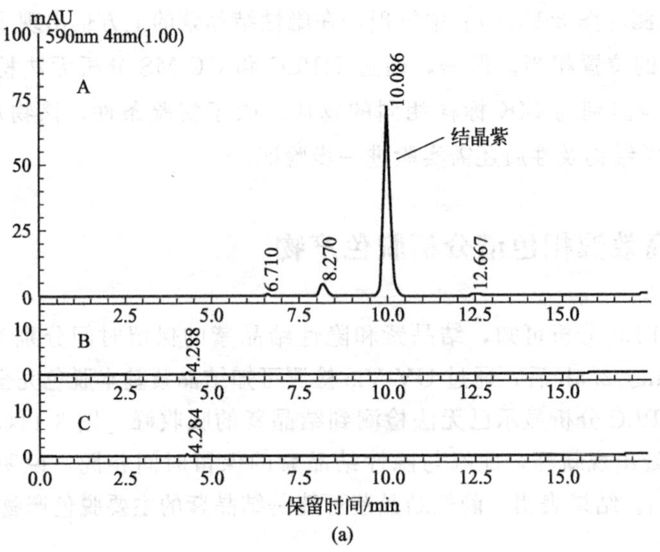

(a)

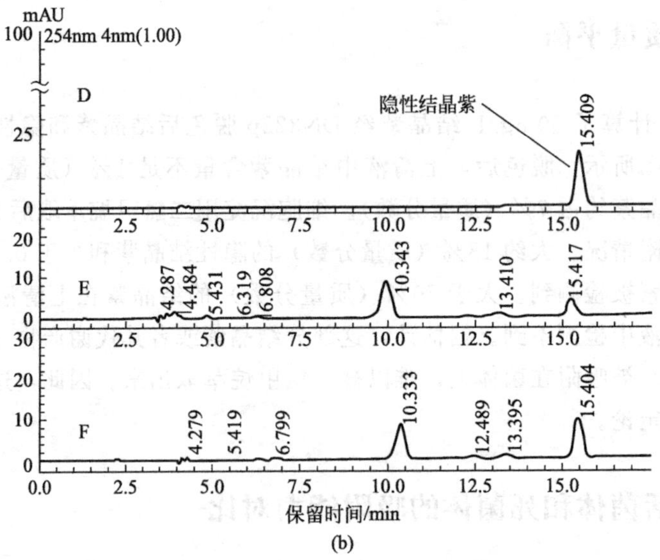

(b)

图 3-14 HPLC 分析结晶紫的 DN322p 脱色产物

(a) 结晶紫检测；(b) 隐性结晶紫检测

A—结晶紫标样；B—上清液二氯甲烷萃取液；C—细胞沉淀二氯甲烷萃取液；D—隐性结晶紫标样；E—上清液二氯甲烷萃取液；F—细胞沉淀二氯甲烷萃取液

结晶紫和隐性结晶紫分别在 590nm 和 254nm 下检测

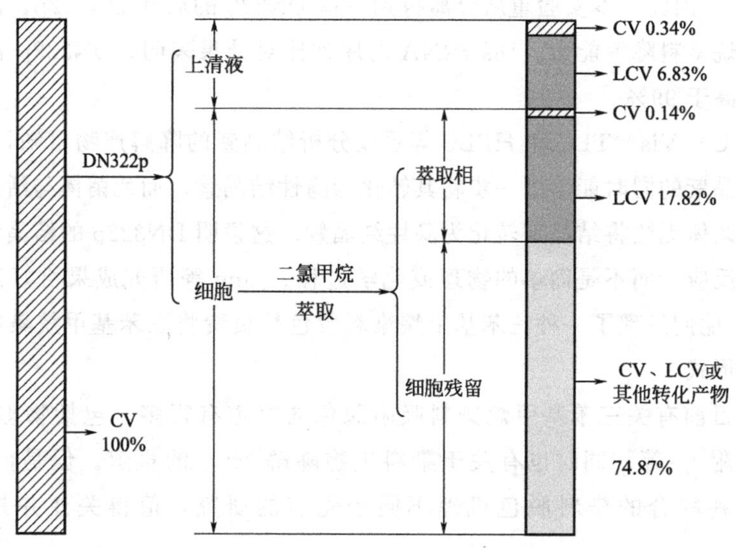

图 3-15　结晶紫脱色后上清液和细胞沉淀中染料相对于初始值的百分含量

CV—结晶紫；LCV—隐性结晶紫

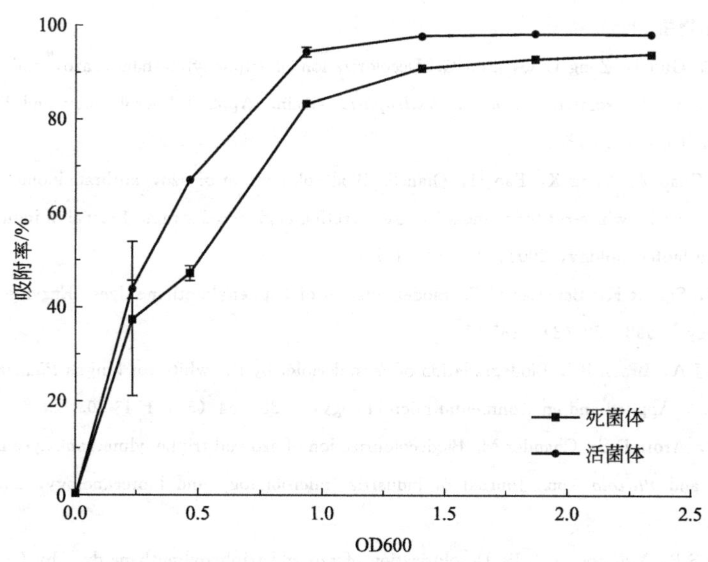

图 3-16　活菌体和死菌体的结晶紫吸附曲线

嗜水气单胞菌 DN322 是一株广谱的染料脱色菌，对很多合成染料都有脱色能力[2]。根据 K. Zengler 的理论，从多样的种群中可以筛选到性状更佳的单细胞后代[29]。因此，本实验重新分离得到一株 DN322 的后代 DN322p，以增强其三苯基甲烷染料降解能力。16s rRNA 的序列比对结果表明，DN322p 和 DN322 的相似性高于 99%。

通过 UV-Vis、TLC 和 HPLC 等手段分析结晶紫的降解产物可知，DN322p 在吸附结晶紫的同时能够进一步将其转化为隐性结晶紫。而死菌体与活菌体不同的是，死菌体无法将结晶紫转化为隐性结晶紫。这说明 DN322p 的结晶紫脱色是一个生物反应，而不是简单的物理或化学过程。Jang 等研究成果恰好说明了这一点[27]。他们发现了一种三苯基甲烷染料脱色酶负责将三苯基甲烷染料转化为它的隐性形式。

尽管目前有关三苯基甲烷染料吸附脱色的报道有很多，包括物理材料[22]和菌体吸附[26]等，同时也有关于染料生物降解[30,31]的报道。但是，DN322p 呈现的这种联合的染料脱色机理不同于先前的研究，值得关注并具有应用前景。

参 考 文 献

[1] 任随周．处理印染废水的 ABR 反应器中微生物生态与脱色菌 DN322 的研究［D］．广州：中国科学院华南植物园．2006.

[2] Ren S Z, Guo J, Zeng G Q, Sun G. Decolorization of triphenylmethane, azo, and anthraquinone dyes by a newly isolated *Aeromonas hydrophila* strain. Applied Microbiology and Biotechnology, 2006, 72 (6): 1316-1321.

[3] Liu W, Chao Y, Yang X, Bao H, Qian S. Biodecolorization of azo, anthraquinonic and triphenylmethane dyes by white-rot fungi and a laccase-secreting engineered strain. Journal of industrial microbiology and biotechnology, 2004, 31 (3): 127-32.

[4] Azmi W, Sani R K, Banerjee U C. Biodegradation of triphenylmethane dyes. Enzyme and microbial technology, 1998, 22 (3): 185-91.

[5] Bumpus J A, Brock B J. Biodegradation of crystal violet by the white rot fungus *Phanerochaete chrysosporium*. Applied and environmental microbiology, 1988, 54 (5): 1143-50.

[6] Gill P K, Arora D S, Chander M. Biodecolourization of azo and triphenylmethane dyes by *Dichomitus squalens* and *Phlebia* spp. Journal of industrial microbiology and biotechnology, 2002, 28 (4): 201-203.

[7] Pointing S B, Vrijmoed L L P. Decolorization of azo and triphenylmethane dyes by *Pycnoporus sanguineus* producing laccase as the sole phenoloxidase. World Journal of Microbiology and Biotechnology, 2000, 16 (3): 317-318.

[8] Vasdev K, Kuhad R C, Saxena R K. Decolorization of triphenylmethane dyes by the birds nest fungus

Cyathus Bulleri. Current Microbiology, 1995, 30 (5): 269-272.

[9] Maier J, Kandelbauer A, Erlacher A, Cavaco-Paulo A, Gübitz G M. A new alkali-thermostable azoreductase from *Bacillus* sp. strain SF. Applied and Environmental Microbiology, 2004, 70 (2): 837-844.

[10] Yatome C, Yamada S, Ogawa T, Matsui M. Degradation of crystal violet by *Nocardia corallina*. Applied Microbiology and Biotechnology, 1993, 38 (4): 565-569.

[11] Yatome C, Ogawa T, Matsui M. Degradation of crystal violet by *Bacillus subtilis*. Journal of Environmental Science and Health Part A, 1991, 26 (1): 75-87.

[12] 任随周，郭俊，王亚丽，岑英华，孙国萍．细菌脱色酶TpmD的酶学特性研究．微生物学报，2006，(05)：823-826.

[13] 任随周，郭俊，王亚丽，岑英华，孙国萍．细菌脱色酶TpmD对三苯基甲烷类染料脱色的酶学特性研究．微生物学报，2006 (3)：385-389.

[14] 王镜岩，朱圣庚，徐长法．生物化学（上册）．北京：高等教育出版社，2002.

[15] 翟中和，王喜忠，丁明孝．细胞生物学．北京：高等教育出版社，2011.

[16] 陈亮，任随周，张培培，许玫英，孙国萍．脱色酶TpmD在大肠杆菌中的高效表达及纯化．中国环境科学，2009，29 (12)：1272-1276.

[17] 卫晋波，孙国萍，任随周，岑英华，王燕．三苯基甲烷类染料氧化酶基因（tpmD）在大肠杆菌中的无诱导表达．环境科学学报，2009，29 (3)：527-535.

[18] 陈亮，任随周，许玫英，孙国萍．乳糖替代IPTG诱导脱色酶TpmD基因在大肠杆菌中的高效表达．微生物学通报，2009，36 (4)：551-556.

[19] 张培培，任随周，许玫英，孙国萍．有机溶剂和抑制剂对毕赤酵母表达TpmD酶活性影响．微生物学报，2009，49 (09)：1190-1195.

[20] 潘涛，刘大伟，任随周，郭俊，孙国萍．DTT对三苯基甲烷染料脱色的研究．环境科学，2012，33 (03)：866-870.

[21] Gregory P. Dyes and dye intermediates//Kroschwitz J I (ed). Encyclopedia of Chemical Technology: Vol 8. New York: John Wiley and Sons, 1993: 544-545.

[22] Kumar R, Ahmad R. Biosorption of hazardous crystal violet dye from aqueous solution onto treated ginger waste (TGW). Desalination, 2011, 265 (1-3): 112-118.

[23] Fan H J, Huang S T, Chung W H, Jan J L, Lin W Y, Chen C C. Degradation pathways of crystal violet by fenton and fenton-like systems: condition optimization and intermediate separation and identification. Journal of Hazardous Materials, 2009, 171 (1): 1032-1044.

[24] Knapp J S, Newby P S, Reece L P. Decolorization of Dyes by Wood-rotting *Basidiomycete* fungi. enzyme and microbial technology, 1995, 17 (7): 664-668.

[25] Nyanhongo G S, Gomes J, Gubitz G M, Zvauya R, Read J, Steiner W. Decolorization of textile dyes by laccases from a newly isolated strain of *Trametes modesta*. Water Research, 2002, 36 (6): 1449-1456.

[26] Jing W, Jung B G, Kim K S, Lee Y C, Sung N C. Isolation and characterization of *Pseudomonas otitidis* WL-13 and its capacity to decolorize triphenylmethane dyes. Journal of Environmental Sciences, 2009, 21 (7): 960-964.

[27] Jang M S, Lee Y M, Kim C H, Lee J H, Kang D W, Kim S J, Lee Y C. Triphenylmethane reductase from *Citrobacter* sp. strain KCTC 18061P: purification, characterization, gene cloning, and overexpression of a functional protein in Escherichia coli. Applied and environmental microbiology, 2005, 71 (12): 7955-7960.

[28] Asad S, Amoozegar M A, Pourbabaee A A, Sarbolouki M N, Dastgheib S M M. Decolorization of textile azo dyes by newly isolated halophilic and halotolerant bacteria. Bioresource Technology, 2007, 98 (11): 2082-2088.

[29] Zengler K. Central role of the cell in microbial ecology. Microbiology and Molecular Biology Reviews, 2009, 73 (4): 712-729.

[30] Chen C H, Chang C F, Ho C H, Tsai T L, Liu S M. Biodegradation of crystal violet by a *Shewanella* sp NTOU1. Chemosphere, 2008, 72 (11): 1712-1720.

[31] Chen C C, Liao H J, Cheng C Y, Yen C Y, Chung Y C. Biodegradation of crystal violet by *Pseudomonas putida*. Biotechnology Letters, 2007, 29 (3): 391-396.

第 4 章
三苯基甲烷染料在浊点系统中的分配

4.1 浊点系统

非离子表面活性剂水溶液在一定的温度下发生浑浊并形成稳定的、透明的两相，此时的两相体系称为浊点系统，相变温度即是浊点。其中，表面活性剂聚集的一相为凝聚层相或表面活性剂富相，表面活性剂浓度较低的一相为水相或稀相。

4.1.1 浊点系统的基本性质

表面活性剂是指具有固定的亲水亲油基团，在溶液的表面能定向排列，并能使表面张力显著下降的物质。表面活性剂的分子结构具有两亲性：一端为亲水基团，另一端为疏水基团。在水溶液中，当表面活性剂的浓度超过临界值时，表面活性剂的分子会自发形成胶束，此时的表面活性剂浓度被称为临界胶束浓度(critical micelle concentration，CMC)。当表面活性剂浓度高于 CMC 时，溶液的表面张力、黏度、吸附性和增溶能力等都会发生显著变化。常用非离子表面活性剂的物理参数如表 4-1 所示[1,2]。

表 4-1 常用非离子表面活性剂的物理参数[1,2]

表面活性剂	疏水基团	EO 单元数	HLB	浊点/℃	CMC/(mmol/L)
聚氧乙二醇单醚，$CH_3(CH_2)_{i-1}O(CH_2CH_2O)_xH$					
Brij 30	十二烷基	4	9.7	分散的	0.023
Emulgen 120	十二烷基	13	—	—	0.2
Brij 35	十二烷基	23	16.9	100	0.09
Emulgen 147	十二烷基	25	—	—	0.1
Brij 52	十六烷基	2	5.3	—	—
Brij 56	十六烷基	10	12.9	64～69	0.0006
Brij 58	十六烷基	20	15.7	—	—
t-辛基苯氧基聚氧乙烯醚，$-(OCH_2CH_2)_xOH$					
Triton X-45	*t*-辛基苯氧基	4.5	9.8	分散的	0.103
Triton X-114	*t*-辛基苯氧基	7.5	12.3	25	0.2

续表

表面活性剂	疏水基团	EO 单元数	HLB	浊点/℃	CMC/(mmol/L)
Triton X-100 (Igepal CA 630)	*t*-辛基苯氧基	9.5	13.4	66	0.21
Triton X-102	*t*-辛基苯氧基	12.5	—	—	0.3
聚氧乙烯壬基苯基醚，$-O(CH_2CH_2O)_xH$					
PONPE 7.5	壬基苯基	7.5	—	5～20	0.085
Igepal CO-610	壬基苯基	8.5	—	—	0.085
壬基苯基	壬基苯基	9.5	—	—	0.08
PONPE 10	壬基苯基	10	—	62～65	0.07～0.085
Tergitol NP-10	壬基苯基	10.5	—	—	0.054
Tergitol TMN 系列，$-(OCH_2CH_2)_xOH$					
Tergitol TMN-3	—	3	8.1	不溶的	—
Tergitol TMN-5	—	5	10.8	不溶的	—
Tergitol TMN-6	—	6	13.1	36	—
Tergitol TMN-10	—	10	—	—	0.16
聚氧乙烯-聚氧丙烯共聚物，(PEO-PPO-PEO)					
Pluronic L 61	聚氧丙烯	—	3	15～19	—
Pluronic L 62	聚氧丙烯	—	7	双浊点	—
Pluronic L 64	聚氧丙烯	—	7～14	＞60	—
Pluronic F 68	聚氧丙烯	—	＞24	＞100	—
Pluronic L 92	聚氧丙烯	—	6	22～26	—

亲水亲油平衡值（Hydrophile-Lyophile Balance，HLB）是衡量表面活性剂极性的重要参数。HLB 值越高，表面活性剂在水中的溶解度越大。具有相同疏水基团的非离子表面活性剂，其 HLB 和浊点环氧乙烷（EO）单元数的增加而增大。相反，具有相同 EO 单元数的非离子表面活性剂，其 HLB 值和浊点随着疏水链的增长而降低。环氧乙烷脂肪醚水溶液的浊点与其疏水链的长短和 EO 单元数有着良好的线性关系[1]。常用乳化剂的 HLB 值见表 4-2。

表 4-2　常用乳化剂的 HLB 值

商品名	化学名	中文名	类型	HLB
Oleic oil	oleic oil	油酸	阴离子	1.0
Span 85/Arlacel 85	sorbitan trioleate	失水山梨醇三油酸酯	非离子	1.8
Atlas G-1706	polyoxyethylene sorbitol beeswax derivative	聚氧乙烯山梨醇蜂蜡衍生物	非离子	2.0

续表

商品名	化学名	中文名	类型	HLB
Span 65/Arlacel 65	sorbitan tristearate	失水山梨醇三硬脂酸酯	非离子	2.1
Atlas G-1050	polyoxyethylene sorbitol hexastearate	聚氧乙烯山梨醇六硬脂酸酯	非离子	2.6
Emcol EO-50	ethyleneglycol fatty acid ester	乙二醇脂肪酸酯	非离子	2.7
Atlas G-1704	polyoxyethylene sorbitol beeswax derivative	聚氧乙烯山梨醇蜂蜡衍生物	非离子	3.0
Emcol PO-50	propylene glycol fatty acid ester	丙二醇脂肪酸酯	非离子	3.4
Atlas G-922	propylene glycol monostearate	丙二醇单硬脂酸酯	非离子	3.4
Atlas G-2158	propylene glycol monostearate	丙二醇单硬脂酸酯	非离子	3.4
Emcol PS-50	propylene glycol fatty acid ester	丙二醇脂肪酸酯	非离子	3.4
Emcol EL-50	ethylene glycol fatty acid ester	乙二醇脂肪酸酯	非离子	3.6
Emcol PP-50	propylene glycol fatty acid ester	丙二醇脂肪酸酯	非离子	3.7
Arlacel C/Arlacel 83	sorbitan sesquioleate	失水山梨醇倍半油酸酯	非离子	3.7
Atlas G-2859	polyoxyethylene sorbitol-4，5-oleate	聚氧乙烯山梨醇-4，5-油酸酯	非离子	3.7
Atmul 67/Atmul 84	glycerol monostearate	单硬脂酸甘油酯	非离子	3.8
Tegin 515/Aldo 33	glycerol monostearate	单硬脂酸甘油酯	非离子	3.8
Ohlan	hydroxylated lanolin	羟基化羊毛脂	非离子	4.0
Arias G-1727	polyoxyethylene sorbitol beeswax derivative	聚氧乙烯山梨醇蜂蜡衍生物	非离子	4.0
Emcol PM-50	propylene glycol fatty acid ester	丙二醇脂肪酸酯	非离子	4.1
Span 80/Arlacel 80	sorbitan monooleate	失水山梨醇单油酸酯	非离子	4.3
Atlas G-917/AtlasG-3851	propylene glycol monolaurate	丙二醇单月桂酸酯	非离子	4.5
Emcol PL-50	propylene glycol fatty acid ester	丙二醇脂肪酸酯	非离子	4.5
Span 60/Arlacel 60	sorbitan monostearate	失水山梨醇单硬脂酸酯	非离子	4.7
Atlas G-2139	diethylene glycol monooleate	二甘醇单油酸酯	非离子	4.7
Atlas G-2146	diethylene glycol monostearate	二甘醇单硬脂酸酯	非离子	4.7
Emcol DO-50/Emcol DS-50	diethylene glycol fatty acid ester	二甘醇脂肪酸酯	非离子	4.7
Atlas G-1702	polyoxyethylene sorbitol beeswax derivative	聚氧乙烯山梨醇蜂蜡衍生物	非离子	5.0
Emcol DP-50	diethylene glycol fatty acid ester	二甘醇脂肪酸酯	非离子	5.1
Aldo 28	glycerol monostearate	单硬脂酸甘油酯	非离子	5.5
Tegin	glycerol monostearate	单硬脂酸甘油酯	非离子	5.5
Emcol DM-50	diethylene glycol fatty acid ester	二甘醇脂肪酸酯	非离子	5.6

续表

商品名	化学名	中文名	类型	HLB
Atlas G-1725	polyoxyethylene sorbitol beeswax derivative	聚氧乙烯山梨醇蜂蜡衍生物	非离子	6.0
Atlas G-2124	diethylene glycol monolaurate	二甘醇单月桂酸酯	非离子	6.1
Emcol DL-50	diethylene glycol fatty acid ester	二甘醇脂肪酸酯	非离子	6.1
Glaurin	diethylene glycol monolaurate	二甘醇单月桂酸酯	非离子	6.5
Span 40/Arlacel 40	sorbitan monopalmitate	失水山梨醇单棕榈酸酯	非离子	6.7
Atlas G-2242	polyoxyethylene dioleate	聚氧乙烯二油酸酯	非离子	7.5
Atlas G-2147	tetraethylene glycol monostearate	四乙二醇单硬脂酸酯	非离子	7.7
Atlas G-2140	tetraethylene glycol monooleate	四乙二醇单油酸酯	非离子	7.7
Atlas G-2800	polyoxypropylene mannitol dioleate	聚氧丙烯甘露醇二油酸酯	非离子	8.0
Atlas G-1493	polyoxyethylene sorbitol lanolin oleate derivative	聚氧乙烯山梨醇羊毛脂油酸衍生物	非离子	8.0
Atlas G-1425	polyoxyethylene sorbitol lanolin derivative	聚氧乙烯山梨醇羊毛脂衍生物	非离子	8.0
Atlas G-3608	polyoxypropylene stearate	聚氧丙烯硬脂酸酯	非离子	8.0
Span 20/Arlacel 20	sorbitan monolaurate	失水山梨醇月桂酸酯	非离子	8.6
Emulphor VN-430	polyoxyethylene fatty acid	聚氧乙烯脂肪酸酯	非离子	8.6
Atbs G-2111	polyoxyethylene oxypropylene oleate	聚氧乙烯氧丙烯油酸酯	非离子	9.0
Atlas G-1734	polyoxythylene sorbitol beeswax derivative	聚氧乙烯山梨醇蜂蜡衍生物	非离子	9.0
Atlas G-2125	tetraethylene glycol monolaurate	四乙二醇单月桂酸酯	非离子	9.4
Brij 30	polyoxyethylene lauryl ether	聚氧乙烯月桂醚	非离子	9.5
Tween 61	polyoxyethylene sorbitan monostearate	聚氧乙烯失水山梨醇单硬脂酸酯	非离子	9.6
Atlas G-2154	hexaethylene glycol monostearate	六乙二醇单硬脂酸酯	非离子	9.6
Tween 81	polyoxyethylene sorbitan monooleate	聚氧乙烯失水山梨醇单油酸酯	非离子	10.0
Atlas G-1218	polyoxyethylene esters of mixed fatty and resin acids	混合脂肪酸和树脂酸的聚氧乙烯酯	非离子	10.2
Atlas G-3806	polyoxyethylene cetyl ether	聚氧乙烯十六烷基醚	非离子	10.3

续表

商品名	化学名	中文名	类型	HLB
Tween 65	polyoxyethylene sorbitan tristearate	聚氧乙烯失水山梨醇三硬脂酸酯	非离子	10.5
Atlas G-3705	polyoxyethylene lauryl ether	聚氧乙烯月桂醚	非离子	10.8
Tween 85	polyoxyethylene sorbitan trioleate	聚氧乙烯失水山梨醇三油酸酯	非离子	11.0
Atlas G-2116	polyoxyethylene oxypropylene oleate	聚氧乙烯氧丙烯油酸酯	非离子	11.0
Atlas G-1790	polyoxyethylene lanolin derivative	聚氧乙烯羊毛脂衍生物	非离子	11.0
Atlas G-2142	polyoxyethylene monooleate	聚氧乙烯单油酸酯	非离子	11.1
Myrj 45	polyoxyethylene monostearate	聚氧乙烯单硬脂酸酯	非离子	11.1
Atlas G-2141	polyoxyethylene monooleate	聚氧乙烯单油酸酯	非离子	11.4
P. E. G. 400 monooleate	polyoxyethylene monooleate	聚氧乙烯单油酸酯	非离子	11.4
Atlas G-2076	polyoxyethylene monopalmitate	聚氧乙烯单棕榈酸酯	非离子	11.6
S-541	polyoxyethylene monostearate	聚氧乙烯单硬脂酸酯	非离子	11.6
P. E. G. 400 monostearate	polyoxyethylene monostearate	聚氧乙烯单硬脂酸酯	非离子	11.6
Atlas G-3300	alkyl aryl sulfonate	烷基芳基磺酸盐	阴离子	11.7
TO1	triethanolamine oleate	三乙醇胺油酸酯	阴离子	12.0
Atlas G-2127	polyoxyethylene monolaurate	聚氧乙烯单月桂酸酯	非离子	12.8
Igepal CA-630	polyoxyethylene alkyl phonol	聚氧乙烯烷基酚	非离子	12.8
Atlas G-1431	polyoxyethylene sorbitol landing derivative	聚氧乙烯山梨醇羊毛脂衍生物	非离子	13.0
Atlas G-1690	polyoxyethylene alkyl aryle ether	聚氧乙烯烷基芳基醚	非离子	13.0
S-307	polyoxyethylene monolaurate	聚氧乙烯单月桂酸酯	非离子	13.1
P. E. G 400 monolurate	polyoxyethylene monolaurate	聚氧乙烯单月桂酸酯	非离子	13.1
Atlas G-2133	polyoxyethylene lauryl ether	聚氧乙烯月桂醚	非离子	13.1
Atlas G-1794	polyoxyethylene castor oil	聚氧乙烯蓖麻油	非离子	13.3
Emulphor EL-719	polyoxyethylene vegetable Oil	聚氧乙烯植物油	非离子	13.3
Tween 21	polyoxyethylene sorbitan monolaurate	聚氧乙烯失水山梨醇单月桂酸酯	非离子	13.3

续表

商品名	化学名	中文名	类型	HLB
Renex 20	polyoxyethylene esters of mixed fatty and resin acide	混合脂肪酸和树脂酸的聚氧乙烯酯	非离子	13.5
Atlas G-1441	polyoxyethylene sorbitol lanolin derivative	聚氧乙烯山梨醇羊毛脂衍生物	非离子	14.0
Atlas G-7596j	polyoxyethylene sotbitan monolaurate	聚氧乙烯失水山梨醇单月桂酸酯	非离子	14.9
Tween 60	polyoxyethylene sorbitan monostearate	聚氧乙烯失水山梨醇单硬脂酸酯	非离子	14.9
Tween 80	polyoxyethylene sorbitan monooleate	聚氧乙烯失水山梨醇单硬脂酸酯	非离子	15.0
Myrj 49	polyoxyethylene monostearate	聚氧乙烯单硬脂酸酯	非离子	15.0
Altlas G-2144	polyoxyethylene monooleate	聚氧乙烯单油酸酯	非离子	15.1
Atlas G-3915	polyoxyethylene oleyl ether	聚氧乙烯油基醚	非离子	15.3
Atlas G-3720	polyoxyethylene stearyl alcohol	聚氧乙烯十八醇	非离子	15.3
Atlas G-3920	polyoxyethylene oleyl alcohol	聚氧乙烯油醇	非离子	15.4
Emulphor ON-870	polyoxyethylene fatty alcohol	聚氧乙烯脂肪醇	非离子	15.4
Atlas G-2079	polyoxyethylene glycol monopalmitate	聚乙二醇单棕榈酸酯	非离子	15.5
Tween 40	polyoxyethylene sorbitan monopalmitate	聚氧乙烯失水山梨醇单棕榈酸酯	非离子	15.6
Atlas G-3820	polyoxyethylene cetyl alcohol	聚氧乙烯十六烷基醇	非离子	15.7
Atlas G-2162	polyoxyethylene oxypropylene stearate	聚氧乙烯氧丙烯硬脂酸酯	非离子	15.7
Atlas G-1741	polyoxyethylene sorbitan lanolin derivative	聚氧乙烯山梨醇羊毛脂衍生物	非离子	16
Myrj 51	polyoxyethylene monostearate	聚氧乙烯单硬脂酸酯	非离子	16
Atlas G-7596P	polyoxyethylene sorbitan monolaurate	聚氧乙烯失水山梨醇单月桂酸酯	非离子	16.3
Atlas G-2129	polyoxyethylene monolaurate	聚氧乙烯单月桂酸酯	非离子	16.3
Atlas G-3930	polyoxyethylene oleyl ether	聚氧乙烯油基醚	非离子	16.6
Tween 20	polyoxyethylene sorbitan monolaurate	聚氧乙烯失水山梨醇单月桂酸酯	非离子	16.7
Brij 35	polyoxyethylene lauryl ether	聚氧乙烯月桂醚	非离子	16.9
Myrj 52	polyoxyethylene (40) stearate	聚氧乙烯（40）硬脂酸酯	非离子	16.9

续表

商品名	化学名	中文名	类型	HLB
Myrj 53	polyoxyethylene stearate	聚氧乙烯硬脂酸酯	非离子	17.9
Pionin D951P	sodium oleate	油酸钠	阴离子	18.0
Atlas G-2159	polyoxyethylene monolaurate	聚氧乙烯单硬脂酸酯	非离子	18.8
Trenamine D	potassium oleate	油酸钾	阴离子	20.0
Atlas G-263	N-cetyl N-ethyl morpholinium ethosulfate	N-十六烷基-N-乙基吗啉基乙基硫酸钠	阳离子	25～30
Texapon K-12	pure sodium lauryl sulfate	纯月桂基硫酸钠	阴离子	40.0

非离子表面活性剂的浊点主要由表面活性剂的化学结构决定[1]。

Triton 系列的非离子表面活性剂由于其浊点较低，使用最为广泛[3-8]。Pluronic 系列产品已经是商品化的嵌段聚合物。嵌段聚合物中环氧丙烷（PO）和 EO 单元的相对疏水性差异使 Pluronic 表现出明显的表面活性剂特征[1]。

4.1.2 有机溶剂对微生物毒性的 lg *P* 规则

有机溶剂的辛醇水分配系数，又称为疏水性指数（lg *P*）的计算公式（4-1）为[1]：

$$\lg P=\lg \frac{\text{有机溶剂在辛醇相的浓度}}{\text{有机溶剂在水相的浓度}} \tag{4-1}$$

lg *P* 反映了有机溶剂疏水性的强弱。通常情况下，lg *P* 越大，有机溶剂的疏水性越强。除实际测定外，有机溶剂的 lg *P* 值可在网站 http：//www. chemicalize. org/查询到。

常用有机溶剂的 lg *P* 值如表 4-3 所示[1,9]。

1987 年，Laane 等发现，有机溶剂的微生物毒性和 lg *P* 值有着极大的相关性[9]。他们通过固定化细胞的厌氧产气量作为有机溶剂抑制微生物生长的响应。考察了介电常数（ε）、有机溶剂的极性（σ）和疏水性指数（lg *P*）三个指标与微生物活性的关系[1]。如图 4-1 所示，结果表明只有 lg *P* 值与微生物活性存在相关性。在中等极性范围内（lg $P<4$），有机溶剂毒性较大；当 lg $P\geqslant 4$ 时，由于有机溶剂在水溶液中的溶解度降低，反而毒性更小。

León 等考察了有机溶剂对细胞的生物相容性（图 4-2）[10]。得到了和 Laane 等相似的结果。在极性有机溶剂 lg $P<2$ 和 lg $P\geqslant 4$ 条件下，有机溶剂的生物相容性较高；而在中等极性（$2\leqslant$ lg $P<4$）范围内，有机溶剂的生物毒性最大。

表 4-3　常用有机溶剂的 lg *P* 值[1,9]

有机溶剂	lg*P*	有机溶剂	lg*P*	有机溶剂	lg*P*
二甲基亚砜	−1.3	间苯二甲酸	1.5	环己烷	3.2
二氧杂环己烷	−1.1	三乙基胺	1.6	苯甲基酮	3.2
N,*N*-二甲基甲酰胺	−1.0	苯乙酸	1.6	苯丙醚	3.2
甲醇	−0.76	乙酸丁酯	1.7	邻苯二甲酸二乙酯	3.3
乙腈	−0.33	氯丙烷	1.8	壬醇	3.4
乙醇	−0.24	苯乙酮	1.8	癸酮	3.4
乙醛	−0.23	己醇	1.8	己烷	3.5
乙酸	−0.23	硝基苯	1.8	丙基苯	3.6
乙氧基乙醇	−0.22	庚酮	1.8	苯甲酸丁酯	3.7
甲酸乙酯	0.16	苯甲酸	1.9	甲基环己烷	3.7
丙醇	0.28	二丙醚	1.9	辛酸乙酯	3.8
丙酸	0.29	己酸	1.9	二戊基醚	3.9
丁酮	0.29	氯仿	2.0	苯甲酸苄酯	3.9
羟基苯乙醇	0.40	苯	2.0	癸醇	4.0
四氢呋喃	0.49	甲基环己醇	2.0	庚烷	4.0
二乙基胺	0.64	苯甲醚	2.1	对异丙基甲苯	4.1
乙酸乙酯	0.68	苯甲酸甲酯	2.2	苯甲酸戊酯	4.2
吡啶	0.71	丙基丁基胺	2.2	二苯醚	4.3
丁醇	0.80	乙酸戊酯	2.2	辛烷	4.5
戊酮	0.80	邻苯二甲酸二甲酯	2.3	十一醇	4.5
丁酸	0.81	辛酮	2.4	癸酸乙酯	4.9
二乙醚	0.85	庚醇	2.4	十二醇	5.0
苯乙醇	0.90	甲苯	2.5	壬烷	5.1
环己酮	0.96	苯甲酸乙酯	2.6	邻苯二甲酸二丁酯	5.4
甲酸丙酯	0.97	苯乙醚	2.6	癸烷	5.6
二烃基苯	1.0	二丁基胺	2.7	十一烷	6.1
甲基丁基胺	1.2	丙酸戊酯	2.7	邻苯二甲酸二苯酯	6.5
乙酸丙酯	1.2	氯苯	2.8	十二烷	6.6
氯乙烷	1.3	辛醇	2.9	邻苯二甲酸二己酯	7.5
戊醇	1.3	壬酮	2.9	十四烷	7.6
环己酮	1.3	二丁基醚	2.9	十六烷	8.8
苯甲醛	1.3	苯乙烯	3.0	邻苯二甲酸二辛酯	9.6
苯乙醇	1.4	四氯甲烷	3.0	油酸丁酯	9.8
环己醇	1.5	戊烷	3.0	邻苯二甲酸二癸酯	11.7
甲基环己酮	1.5	乙基苯	3.0	邻苯二甲酸二月桂酸酯	13.7
苯酚	1.5	二甲苯	3.0		

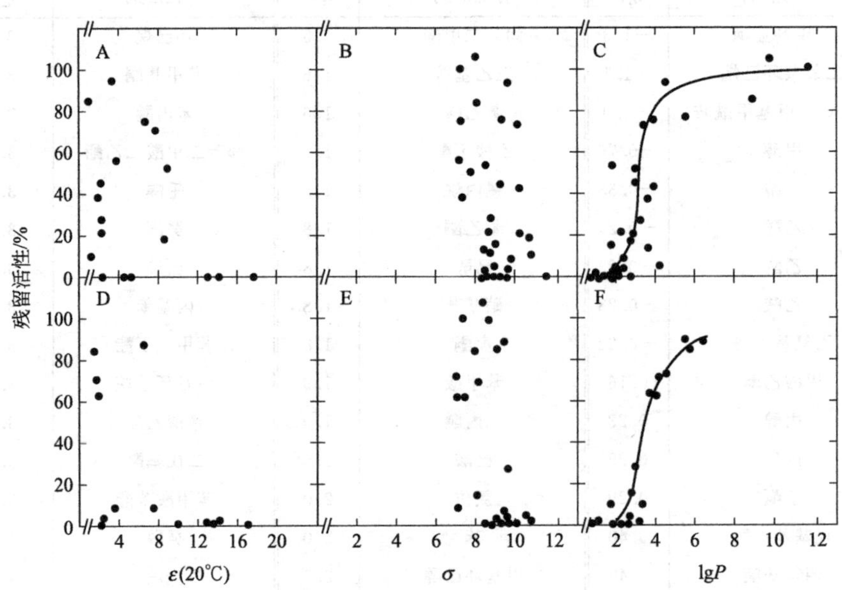

图 4-1　有机溶剂物理化学参数与毒性的关系[9]

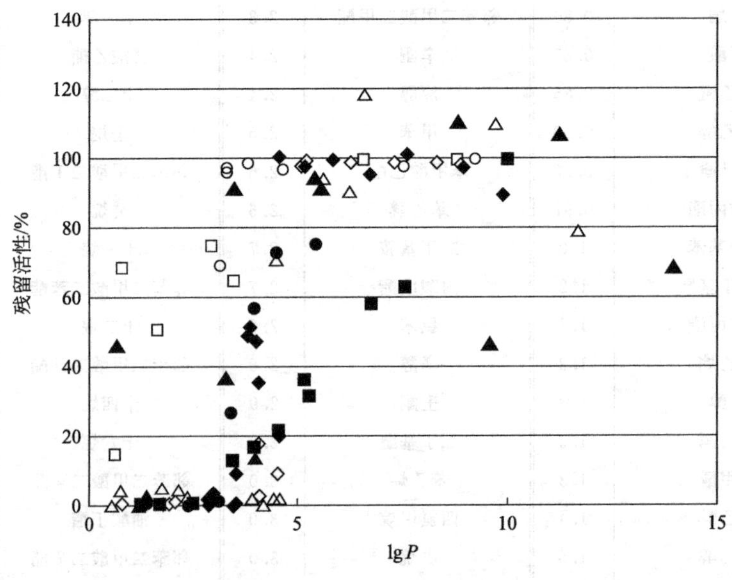

图 4-2　有机溶剂疏水性与细胞活性的关系[10]

4.1.3 浊点系统中物质分配的 lg P 准则

浊点系统在萃取分离中的应用广泛[11,12]，包括萃取金属离子[13,14]、提取发酵产物[15,16]、酶纯化[17]、痕量分析[18-20]以及环境污染物的治理[8,21-25]等。Wang 等将酚类化合物的 lg P 值与其在浊点系统中的分配关联起来，总结并预测了酚类化合物在浊点系统中的分配规律[26]。如图 4-3 所示，随着酚类化合物 lg P 值的增大，其在浊点系统中的分配效率越高[26]。

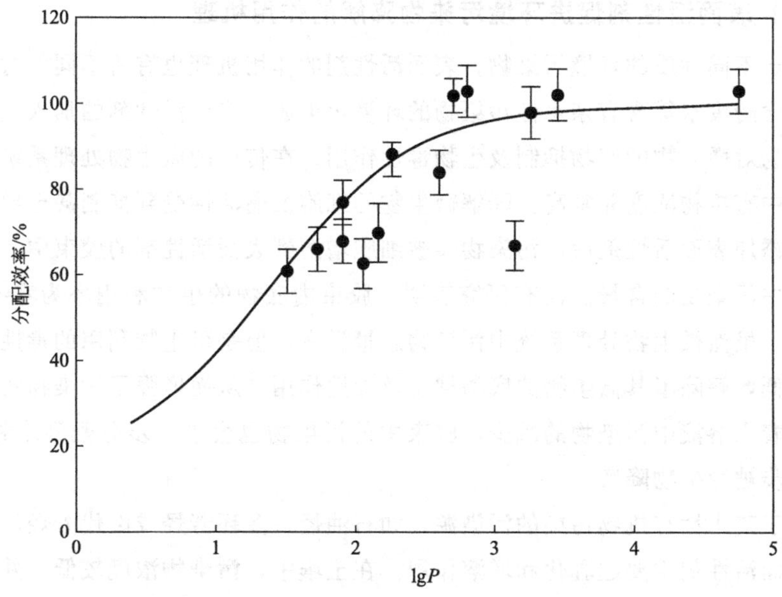

图 4-3 酚类化学物在浊点系统中的分配规律[26]

4.1.4 表面活性剂在环境污染治理中的应用

表面活性剂可分为阴离子型、阳离子型、两性型和非极性型。环境中常用的表面活性剂有 SDS（阴离子型）、苯扎溴铵（阳离子型）、卵磷脂（两性型）以及吐温类（非离子型）表面活性剂等。在环境污染物的生物降解方面，非离子型表面活性剂的应用最广泛[27-30]。化学结构的研究表明，正电荷阳离子的表面活性剂毒性最强，一般用于杀菌剂。阴离子型表面活性剂的毒性稍弱一些，并且对革兰氏阳性菌比对革兰氏阴性菌的毒性更强。一般情况下，非离子型表面活性剂被认为是无毒的。

近年来，有关生物表面活性剂在环境中应用的报道越来越多[31]。很多种微生物都可产生物表面活性剂。生物表面活性剂种类繁多，结构多样，一般可分为：a. 糖脂；b. 脂肽；c. 脂肪酸、中性脂质、磷脂；d. 聚合物；e. 微粒。一般来说，在工业化应用的时候，人们普遍认为生物表面活性剂更加低毒和环境友好。然而，在自然界中，微生物分泌的生物表面活性剂一般用来抵抗外部环境，通过杀菌作用等形成有利于自身在微生物群体中的竞争性优势（例如，偏害共栖）。因此，生物表面活性剂在环境中的原位应用，要充分考虑到其对环境微生物种群的影响。

(1) 表面活性剂促进环境污染物降解的作用机理

对于不同性质的环境污染物，表面活性剂的作用机理也有所不同。对于染料废水、含酚废水等含有亲水性污染物的环境污染源，表面活性剂的加入主要是解除污染物对微生物的底物抑制及生物毒性作用。在传统的微生物处理系统中，由于废水中有机物浓度非常高，降解微生物的细胞及酶活性受到抑制甚至导致细胞凋亡。添加表面活性剂后，污染物会被细胞增溶到表面活性剂的胶束中，降低了水溶液中污染物的含量。而有研究表明，胶束内底物的生物利用率为零[32]。这样一来，虽然微生物处理系统中污染物总量没变，但是可生物利用的毒性污染物浓度降低，解除了其微生物的底物抑制及毒性作用，从而增强了污染物的生物降解。随着水溶液中污染物的减少，胶束中的污染物也会进一步分配到水相当中，可进一步被微生物降解。

对于疏水性有机物污染的污染源，如石油烃、多环芳烃或卤代芳烃污染的土壤，表面活性剂主要起乳化和增溶作用。在土壤中，污染物浓度较低，并且牢固地吸附在土壤颗粒表面和内部，难溶于土壤内部的水环境中。这直接限制了污染物的生物利用度。在污染土壤中添加表面活性剂，通过乳化和增溶作用，可提高污染物在土壤水环境中的溶解度，加快污染物在土壤颗粒、表面活性剂胶束和水环境之间的传质作用。增溶到水环境中的污染物更容易被微生物利用。

另外，表面活性剂对降解微生物的生物活性也有影响。微生物的细胞膜是由磷脂双分子层组成，在水溶液中加入表面活性剂，微生物细胞膜受到表面活性剂的吸附作用，其通透性及流动性增强。污染物更容易跨膜进入微生物细胞内，有利于加快污染物降解速率。

(2) 表面活性剂在环境污染物生物降解中的应用范围

表面活性剂在环境污染物降解中的应用十分广泛。根据目标污染物的不同，可分为以下几大类。

① 染料。染料废水的环境污染问题一直受到人们的普遍关注。其废水中含有大量的有机物和盐分，具有COD高、色泽深、酸碱性强等特点，一直是废水处理中的难题。Saxena等研究了Triton X-100、Tween 80和SDS对白腐真菌去除纸浆废水中颜色、木质素以及COD的作用，发现Tween 80的作用效果最显著。添加100mg/L的Tween 80可将颜色、COD和木质素的去除率分别从34.49%、40.74%和16.38%提高到81.29%、75.35%和65.84%[33]，效果显著。Avramova等详细对比了非离子表面活性剂TX-100、阳离子表面活性剂十六烷基三甲基溴化铵（CTAB）及阴离子表面活性剂月桂酰肌氨酸钠（SLS）对微生物脱色酸性橙Ⅰ（acid orange 7，AO7）的效果。结果发现，TX-100和SLS会抑制染料的脱色，但CTAB可显著提高染料脱色率[34]。这可能与阳离子型表面活性剂与细胞间的电位差有关。CTAB可将染料牢固地结合在微生物细胞上，加快了底物的传递作用，提高了染料的生物利用度。Gül等也发现，阴离子型表面活性剂十二烷基苯磺酸盐（SDBS）会抑制雷玛唑亮蓝（Remazol brilliant blue）的生物脱色，而十二烷基三甲基溴化铵（DTAB）可显著增强微生物的染料脱色作用[27]。

② 脂肪烃。在石油工业、天然气工业区附近的土壤及水体中，一般存在严重的脂肪烃污染。脂肪烃不溶于水，形成非水相，生物利用度低。表面活性剂的加入可将脂肪烃乳化，提高其生物利用度。在水溶液中添加300mg/L的鼠李糖脂，十八烷在水中的分散浓度提高了4个数量级。1500mg/L的十八烷84h后降解了20%，而未添加表面活性剂的对照样品只有5%被降解[35]。Churchill等发现，表面活性剂增强的生物降解作用似乎与降解微生物的细胞疏水性有关。添加TX-100和Tween 80后，亲水的绿脓假单胞菌ATCC 9027和红平红球菌的十八烷降解速率增强，而疏水的对不动杆菌的十八烷生物降解不起作用[36]。Owsianiak等的研究成果却得到了不同的结论。他们发现，离子型的表面活性剂鼠李糖脂和非离子型的表面活性剂TX-100对微生物降解柴油的影响似乎只与菌种特异性有关，与细胞表面疏水性和表面活性剂类型无关[37]。因此，表面活性剂类型、细胞疏水性是否影响到脂肪烃的生物降解，还需研究者进一步探索。

③ 单环芳烃。单环芳烃污染物一般包括苯酚、甲苯和二甲苯等，主要存在于焦化废水和工业废气当中，可污染地下水、河流和大气。Ding等通过添加TX-100、CTAB、Tween 80和鼠李糖脂强化热带假丝酵母CICC 1463降解苯酚，结果发现CTAB具有生物毒性，抑制了苯酚的降解。低浓度的Tween 80、TX-100和鼠李糖脂可促进苯酚降解，并且微生物可利用鼠李糖脂作为辅底物[38,39]。Chan等研究了Brij 30和Brij 35在生物滴滤池中对甲苯生物降解的作用，发现

表面活性剂的加入增强了甲苯在水中的溶解，但抑制了甲苯的生物降解[40,41]。Inakollu 等通过加入两种鼠李糖脂生物表面活性剂，将苯和甲苯的降解速率常数分别提高了近 25％和 27％[42]。

④多环芳烃。环境中的多环芳烃主要来源于煤和石油的燃烧。多环芳烃大多吸附在大气和水中的微小颗粒物上。大气中的多环芳烃又可通过沉降和降水冲洗作用而污染土壤和地面水。通常情况下，多环芳烃的环越多生物降解越难。目前，对表面活性剂影响多环芳烃微生物降解的研究比较多。1995 年，大多数研究者们都在探索表面活性剂在多环芳烃生物降解过程中是否会产生积极的作用。Deschenes 等发现阴离子表面活性剂 SDS 可以增溶四环以下的多环芳烃，但抑制了其生物降解[43]。Liu 等发现在萘微生物降解过程中，添加非离子表面活性剂 Brij 30 和 TX-100 不会抑制菌的生长和萘的降解[44]。Tsomides 等测试了 15 种表面活性剂对菲生物降解的影响，发现除 TX-100 外，其他表面活性剂都对多环芳烃降解菌有毒性作用[45]。由于表面活性剂在多环芳烃生物降解中的不确定性，1996 年，Grimberg 等建立了菲在非离子表面活性剂 Tergitol NP-10 中的增溶及生物降解模型，采用的菌种为施氏假单胞菌 P16（*Pseudomonas stutzeri* P16）[29]。此模型对多环芳烃在表面活性剂胶束水溶液中的生物利用度有一定的指导作用。随后，表面活性剂增强多环芳烃生物降解的报道越来越多。采用的表面活性剂多为非离子型的，如 Triton X-100[46-49]、Tergitol NP-10[46]、Igepal 系列[46]、Tween 系列[47,50-53]、Brij 系列[54-57]及生物表面活性剂鼠李糖脂[58]等。

⑤ 卤代芳烃。滴滴涕（dichlorodiphenyltrichloroethane，DDT）是一种有机氯类杀虫剂，属神经及实质脏器毒物，对人和大多数其他生物体具有中等强度的急性毒性。早在 1989 年，Kile 等就研究了表面活性剂 TX-100、TX-114、TX-405、Brij 35、SDS 和 CTAB 对 DDT 的增溶能力[59]。1996 年，You 等发现表面活性剂可增强 DDT 的厌氧生物转化[60]。随后，Aislabie 等详细总结了 DDT 的微生物转化及表面活性剂的作用[61]。他指出，由于 DDT 在土壤中的生物利用度很低，用表面活性剂预处理污染土壤是必须的步骤。Baczynski 等加入 0.5mmol/L Tween 80 可将 DDT 的转化率提高 2 倍[28]。升高表面活性剂浓度到 1.25mmol/L，DDT 的转化效率却下降到与对照样品同等水平，并且不利用产物 4,4-二氯二苯甲酮(4,4′-dichlorobenzophenone，DBP）的形成。

多氯联苯是典型的卤代芳烃污染物。多氯联苯极难溶于水而易溶于脂肪和有机溶剂，并且极难分解，因而能够在生物体脂肪中大量富集，有很强的生物毒害作用。多氯联苯的化学性质非常稳定，很难在自然界分解，属于持久性的有机污染物。1998 年，Fava 等研究了在多氯联苯污染土壤异位生物修复过程中，表面活性剂的作用。发现在泥浆状态时，TX-100 抑制了多氯联苯的生物利用度，而皂树（quillaya saponin）可轻微地增强多氯联苯的生物降解；在固定床反应器中，TX-100 不会抑制微生物活性并轻微地增强了多氯联苯的生物降解，而皂树对多氯联苯的生物降解没有影响[30]。Rojas-Avelizapa 等分析了三种非离子表面活性剂 Tween 80、Tergitol NP-10 和 TX-100 对多氯联苯生物降解的影响，Tween 80 效果最佳，多氯联苯的降解率达到 39%～60%[62]。一些生物表面活性剂不仅能够增溶多氯联苯，还可作为降解菌的碳源被生物利用，可避免表面活性剂的二次污染。Singer 等利用失水山梨醇三油酸酯作为表面活性剂和降解菌的唯一碳源，增强二氯联苯的降解[63]。

多溴联苯醚（poly brominated diphenyl ethers，PBDES）是一类环境中广泛存在的全球性有机污染物。由于其具有环境持久性、远距离传输性、生物可累积性及对生物和人体具有毒害效应等特性，对其环境问题的研究已成为当前环境科学的一大热点[64,65]。PBDES 的微生物降解还处在起始阶段，相关报道不多[66]。其中表面活性剂对 PBDES 微生物降解的强化作用只有国内南京大学的高士祥教授团队做了相关研究，发现 Tween 80 和 β-环糊精可增强十溴联苯醚 BDE-209 的好氧生物降解，该研究处于国际领先水平[67,68]。

⑥ 表面活性剂的生物降解。尽管表面活性剂在环境污染物生物降解中的作用显著，但其是否会带来二次污染也是其规模应用所必须要考虑的问题。一些好氧和厌氧的实验表明，环境常用的表面活性剂如 Tween 80 和 TX-100 等，可以在自然环境中快速降解[69]。而且，对于乙氧基非离子表面活性剂，降低亲水基团中氧乙烯单位的数量和增加疏水基团的线性度都会强化其生物降解[70]。

有些表面活性剂及其代谢产物是有毒的，这类表面活性剂通常情况下应避免在环境中使用。如烷基酚羟乙基物表面活性剂在美国和几个其他国家被禁止使用，因为这些聚合物长链切断以后会增强疏水性和毒性[70]。而且，烷基酚羟乙基物被报道会降解成顽固代谢产物，这种产物表现出对女性雌激素的生物活性。如果摄入高剂量可能会导致癌症或者其他健康问题[70]。

生物表面活性剂，如环糊精和鼠李糖脂等，比商业合成的表面活性剂更容易降解，不会产生二次污染。尽管生物表面活性剂或者改良环糊精对于环境污染的修复前途广阔，但是表面活性剂的生物合成还属于新兴技术[31]，其规模化工业应用受到成本的限制，还有很长的一段路要走。

4.2 三苯基甲烷染料的浊点萃取过程

第 3 章主要阐述了嗜水气单胞菌 DN322p 的结晶紫脱色特性。然而，由于脱色后产生的隐性结晶紫仍然具有生物毒性[71]，因此找到一种可以彻底脱色脱毒的三苯基甲烷染料生物降解体系显得尤为重要。

本章拟运用介质工程的原理[1]，将非离子表面活性剂水溶液形成的浊点系统用于三苯基甲烷染料的生物降解体系，以期提高脱色效率并有效脱毒。为达成这一目标，清楚地了解三苯基甲烷染料在浊点系统中的分配规律是必要前体。本章致力于探索三苯基甲烷染料在浊点系统中的分配行为。

在分析化学中，浊点系统常用于分离金属离子、有机物和生物活性物质，这一过程叫做浊点萃取[11,72]。当前，浊点萃取在环境废水处理中的应用发展快速。例如分离毒性曙红[25]、苯[22]、直接黄[23]、罗丹明 B[8] 和橘红[73] 等。然而，有关浊点萃取三苯基甲烷染料的报道还很有限。Pourreza 等利用非离子表面活性剂 Triton X-100 在 60℃下去除染料孔雀石绿[74]。An 等利用浊点萃取浓缩并分析水样中的微量孔雀石绿和结晶紫[19]。

为了使浊点萃取过程更加经济，有机物必须被反萃出表面活性剂以达到重复利用的目的。但由于非离子表面活性剂的非挥发性，一方面被称为绿色溶剂，一方面又难以像传统的有机溶剂那样可以蒸发回收[75]。在某些情况下，Winsor 微乳液可以用于回收非离子表面活性剂[76]。Wang 等利用 Wisnor Ⅰ微乳液成功从 PEG 诱导的浊点系统中分离出 L-PAC 并回收了 Triton X-100[77,78]。本人通过异丁醇和 2-苯乙醇萃取的方法，利用 Winsor Ⅱ微乳液成功回收了 Triton X-114 和 Triton X-45 混合表面活性剂[79,80]。然而，如何从浊点系统凝聚层相中分离染料并回收表面活性剂的方法还未见报道，需要进一步探索。

本实验中，采用浊点系统萃取四种三苯基甲烷染料，包括结晶紫、乙基紫、孔雀石绿和灿烂绿。详细研究了温度、表面活性剂浓度和染料浓度对四种染料分配效率的影响。采用二氯甲烷萃取的方法回收稀相中的表面活性剂。并且，利用 Langmuir 吸附曲线拟合萃取过程。通过对比分析染料最大吸附量、lg P 和染料分子结构三者之间的关系后得知，三苯基甲烷染料在浊点系统中的增溶与其分子结构和 lg P 值相关。

4.2.1 染料与非离子表面活性剂储备液

分别称取 0.408g、0.464g、0.492g 和 0.482g 的结晶紫、乙基紫、孔雀石绿和灿烂绿溶解在 200mL 蒸馏水中，制成 5mmol/L 的储备液。称取 107.4g 的 Triton X-114 溶于水中得到 0.2mol/L 的表面活性剂储备液。更多稀释的染料和表面活性剂溶液用这些储备液稀释后使用。几种染料的基本性质如表 4-4 所示。

表 4-4 四种三苯基甲烷染料结构及其疏水性指数

染料	结构	lg P①
结晶紫	H_3C $\overset{+}{N}$ CH_3; Cl^-; H_3C N CH_3; CH_3 N CH_3	1.77
孔雀石绿	Cl^-; H_3C N CH_3; N^+ CH_3 CH_3	1.69
乙基紫	H_3C $\overset{+}{N}$ CH_3; Cl^-; CH_3 N CH_3; CH_3 N H_3C	3.92
灿烂绿	H_3C $\overset{+}{N}$ CH_3; HO—S(=O)(=O)—O^-; N CH_3 H_3C	3.13

续表

染料	结构	lg P①
灿烂绿	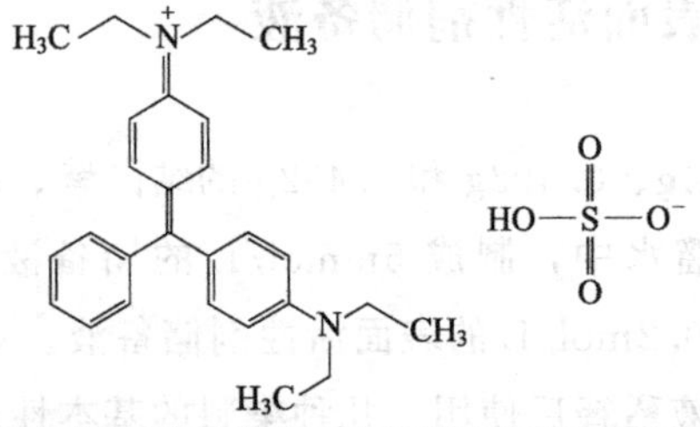 H_3C、N^+、CH_3、HO—S(=O)(=O)—O^-、N、CH_3、H_3C	3.13

① lg P 值获取自数据库 Chemicalize database（http：//www. chemicalize. org/）。

4.2.2 染料的浊点萃取步骤

将不同浓度的非离子表面活性剂 Triton X-114、染料和盐的水溶液置于10mL 的刻度玻璃离心管中，水浴条件下静置 2h 以上。彻底分相后，记录凝聚层相体积。取出部分稀相分析染料和表面活性剂的浓度。

用二氯甲烷回收稀相中的 Triton X-114。将不同体积比的二氯甲烷和稀相水溶液在 1.5mL 离心管中混合并在 10000r/min 下离心分离 5min。取出水相分析 Triton X-114 的浓度。

实验中，结晶紫的浓度为 25μmol/L、50μmol/L、100μmol/L 和 200μmol/L、孔雀石绿的浓度为 50μmol/L、100μmol/L、200μmol/L 和 400μmol/L、乙基紫和灿烂绿的浓度为 100μmol/L、200μmol/L、400μmol/L 和 500μmol/L。Triton X-114 的浓度为 0.05～0.2mol/L。温度选取在 40℃、45℃、50℃和 55℃。NaCl 和 $CaCl_2$ 用于测试盐对染料在浊点系统中分配的影响。盐的浓度选取在 0.05mol/L、0.1mol/L、0.2mol/L、0.3mol/L、0.4mol/L 和 0.5mol/L。

根据染料浊点萃取的结果，配制 0.19mol/L 和 1.86mol/L 的 Triton X-114 水溶液代表稀相中表面活性剂的最低和最高浓度。二氯甲烷与稀相的体积比在 0.05～1.0 之间变化。

4.2.3 表面活性剂与染料浓度分析方法

用分光光度计检测染料和表面活性剂的浓度。纯的 Triton X-114、结晶紫、乙基紫、孔雀石绿和灿烂绿在通过全波长扫面后得知其最大吸收峰分别在 224nm、582nm、616nm、595nm 和 625nm。表面活性剂 Triton X-114 不会显著影响染料的最大吸收。设定一定的浓度梯度后，测定几种物质的标准曲线用于计算浓度。

对于浊点萃取，染料的萃取率定义为式(4-2)：

$$\text{染料的萃取率}=\left(1-\frac{\text{稀相中的染料浓度}}{\text{染料初始浓度}}\right)\times 100\% \tag{4-2}$$

凝聚层相的体积分数按式(4-3) 计算：

$$\text{凝聚层相的体积分数}=\frac{\text{凝聚层相的体积}}{\text{溶液总体积}} \tag{4-3}$$

对于溶剂萃取，Triton X-114 的回收率如式(4-4) 所示：

$$\text{Triton X-114 的回收率}=\left(1-\frac{\text{溶剂萃取后水相中 Triton X-114 的浓度}}{\text{稀相中 Triton X-114 的初始浓度}}\right)\times 100\% \tag{4-4}$$

4.3 三苯基甲烷染料的浊点萃取特性与增溶平衡

4.3.1 染料浓度对浊点的影响

浊点是浊点萃取的一个重要参数，它决定了浊点萃取的操作温度。为确保实验过程中操作温度高于浊点，本实验详细测定了染料对 Triton X-114 水溶液浊点的影响，浓度为 10g/L 的 Triton X-114 水溶液在常压下浊点为 25℃[6]。在单独存在一种三苯基甲烷染料的情况下，Triton X-114 的浊点有所提高（图 4-4）。如图 4-4 所示，孔雀石绿在浓度为 100μmol/L、200μmol/L 和 400μmol/L 时，系统的浊点分别提高到 28.5℃、41℃和 51℃。对于结晶紫和乙基紫，在与孔雀石绿相同的浓度下，浊点分别被提高到 32℃和 28℃、43.5℃和 46℃、93℃和 95℃。对于灿烂绿，浊点变化最显著。当灿烂绿的浓度从 100μmol/L 提高到 200μmol/L 时，系统的浊点从 32℃提高到 93℃。

4.3.2 浓度和温度对染料分配行为的影响

浊点和操作温度的差异影响着胶束的结构和物质在胶束中的增溶[75]。因此，

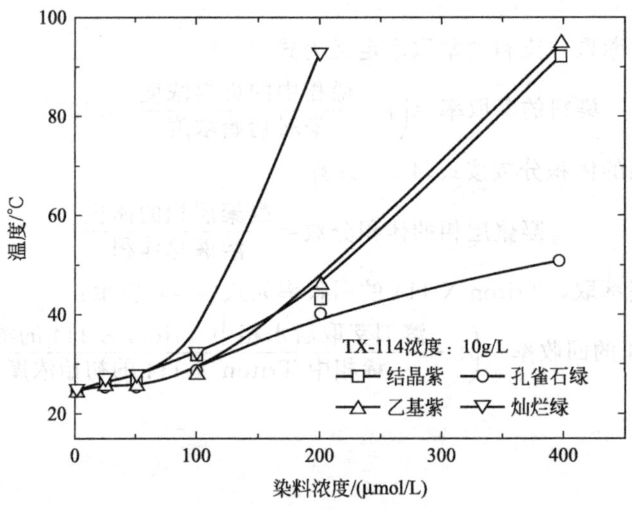

图 4-4　不同染料对浊点的影响

如图 4-5、图 4-6、图 4-7 和图 4-8 所示，在不同的 Triton X-114 浓度、染料浓度和操作温度下，测定了四种三苯基甲烷染料在浊点系统中的分配效率。在图 4-5

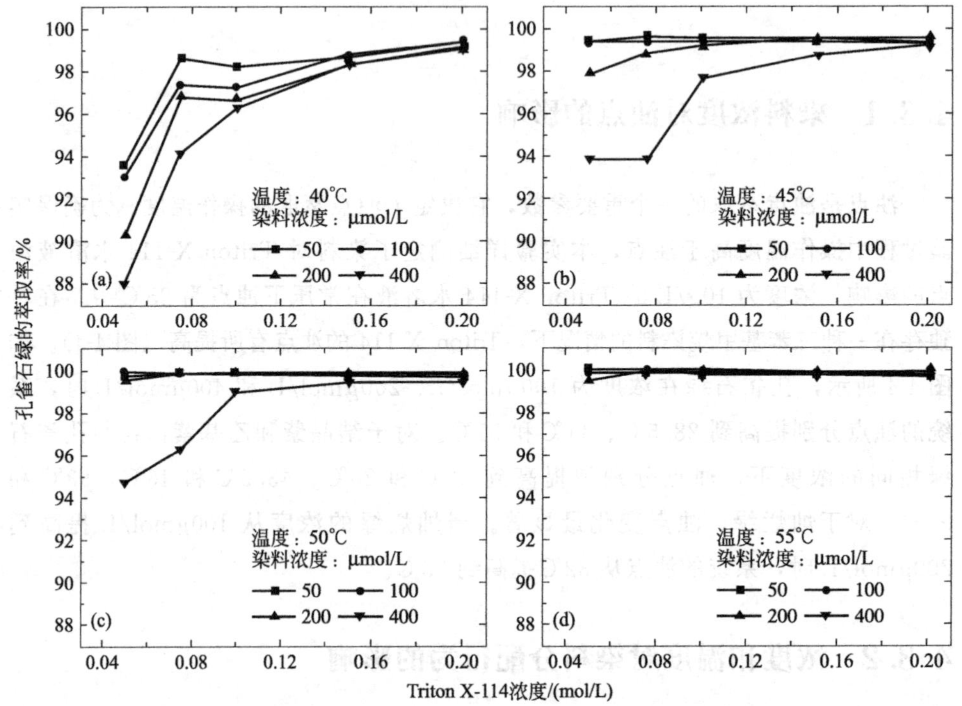

图 4-5　不同温度 Triton X-114 和孔雀石绿浓度对孔雀石绿萃取率的影响

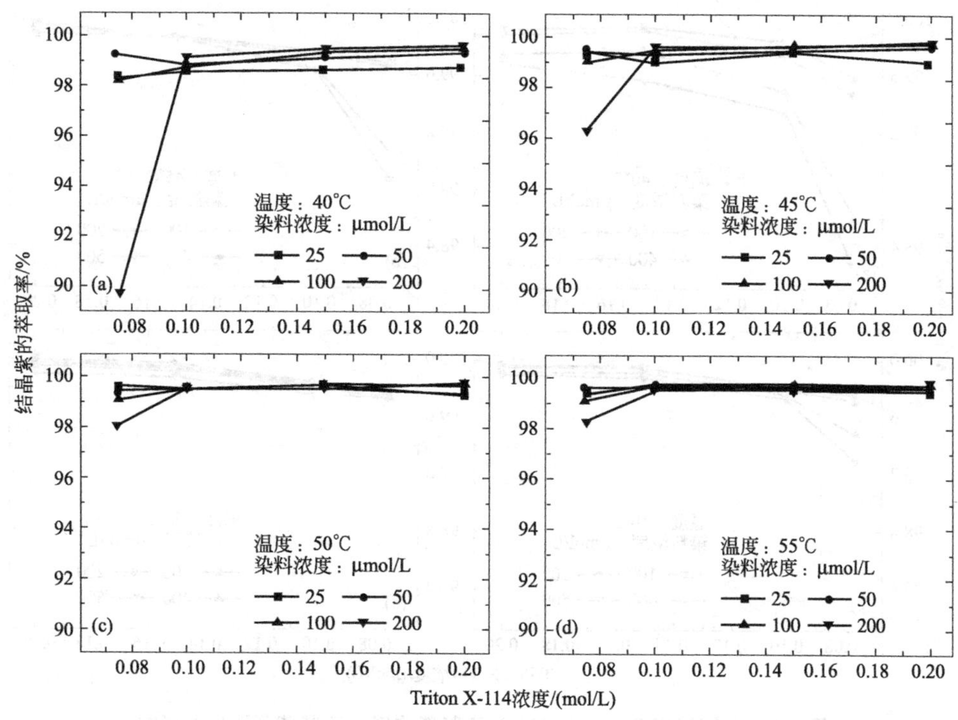

图 4-6　不同温度 Triton X-114 和结晶紫浓度对结晶紫萃取率的影响

(a) 中，当 Triton X-114 的浓度从 0.05mol/L 提高到 0.075mol/L 时，孔雀石绿的萃取率显著增大。Triton X-114 的浓度高于 0.075mol/L 后，染料萃取率增大减缓。对于 50μmol/L 的孔雀石绿，当 Triton X-114 的浓度为 0.075mol/L 时，萃取率达到 98%。对于 100μmol/L、200μmol/L 和 400μmol/L 的孔雀石绿，大约需要 0.15mol/L 的 Triton X-114 才能达到相同的萃取效果。

如图 4-5(a) 所示，在一定的染料浓度和温度下，孔雀石绿的萃取率随 Triton X-114 浓度的增大而增大。通常情况下，胶束浓度随表面活性剂浓度的增大而增大，进一步导致胶束中增溶更多染料[75]。由于稀相中胶束浓度接近 CMC[75]，所以增加表面活性剂浓度可形成更多凝聚层相来萃取染料。因此，增加表面活性剂浓度可提高染料的萃取率。图 4-5(a) 中，染料的萃取率随染料浓度的升高而降低。在固定的操作温度和表面活性剂浓度下，染料在凝聚层相和稀相中的分配率是恒定的[23]。所以，浊点系统凝聚层相的染料增溶能力是一定的。当增大染料浓度时，更多无法被萃取的染料残留在稀相。

如图 4-5 所示，操作温度对孔雀石绿在浊点系统中的分配产生影响。染料萃

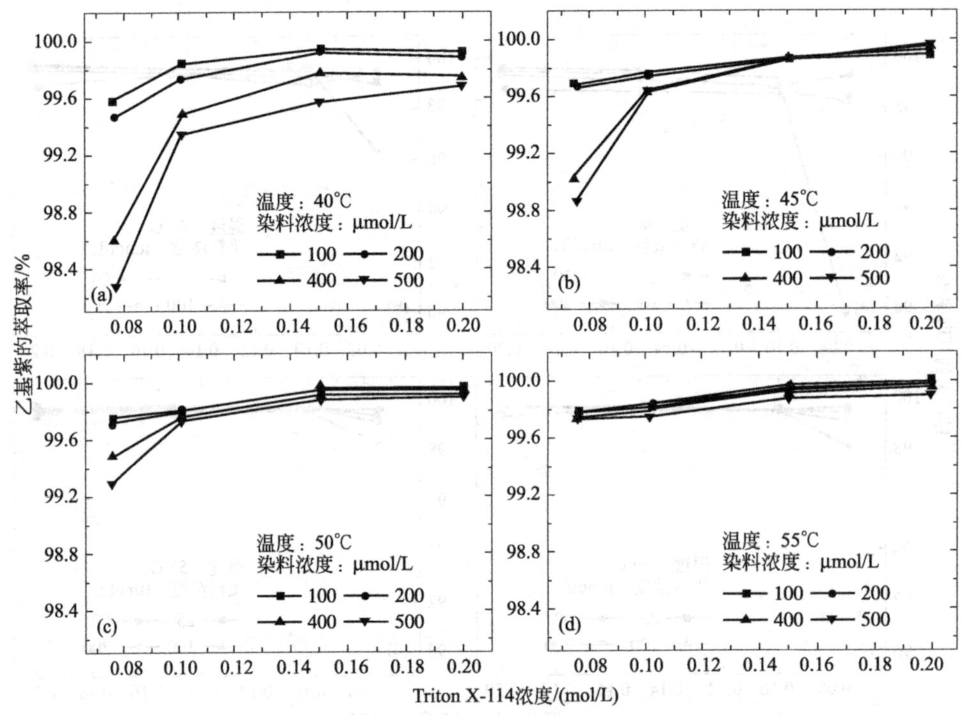

图 4-7 不同温度 Triton X-114 和乙基紫浓度对乙基紫萃取率的影响

取率随温度的升高而增大。例如，在 400μmol/L 孔雀石绿和 0.05mol/L Triton X-114 的情况下，染料的萃取率从 40℃时的 88%［图 4-5(a)］提高到了 55℃时的 99%［图 4-5(d)］。对于结晶紫、乙基紫和灿烂绿，温度对其萃取的影响具有相似的趋势（图 4-6、图 4-7 和图 4-8）。例如，在 200μmol/L 结晶紫和 0.05mol/L Triton X-114 的情况下，染料的萃取率在 40℃时约为 90%［图 4-6(a)］，45℃时约为 96%［图 4-6(b)］，55℃时约为 98%［图 4-6(d)］。随着温度的升高，浊点系统的疏水性逐渐增强。相应地，胶束聚集并且脱水[11,72]，这使得凝聚层相的增溶能力更强。因此，随着更多的染料分子增溶到凝聚层相当中，染料的萃取率也随温度的升高而增大。

当然，一旦染料的萃取率达到一定高的水平，进一步通过升高温度来提高染料的萃取率将变得困难。例如，50℃时，灿烂绿的萃取率为 99%［图 4-8(c)］，进一步提高温度到 55℃时，染料萃取率并未发生明显变化［图 4-5(d)］。值得一提的是，400μmol/L 和 500μmol/L 乙基紫在 Triton X-114浓度为 0.05mol/L 的浊点系统中，当温度达到 55℃时，溶液分相形成三相不稳定体系：2 个凝聚层相和 1 个稀相。染料萃取率也急剧下降。在这种状态下，浊点系统已经不是传统意

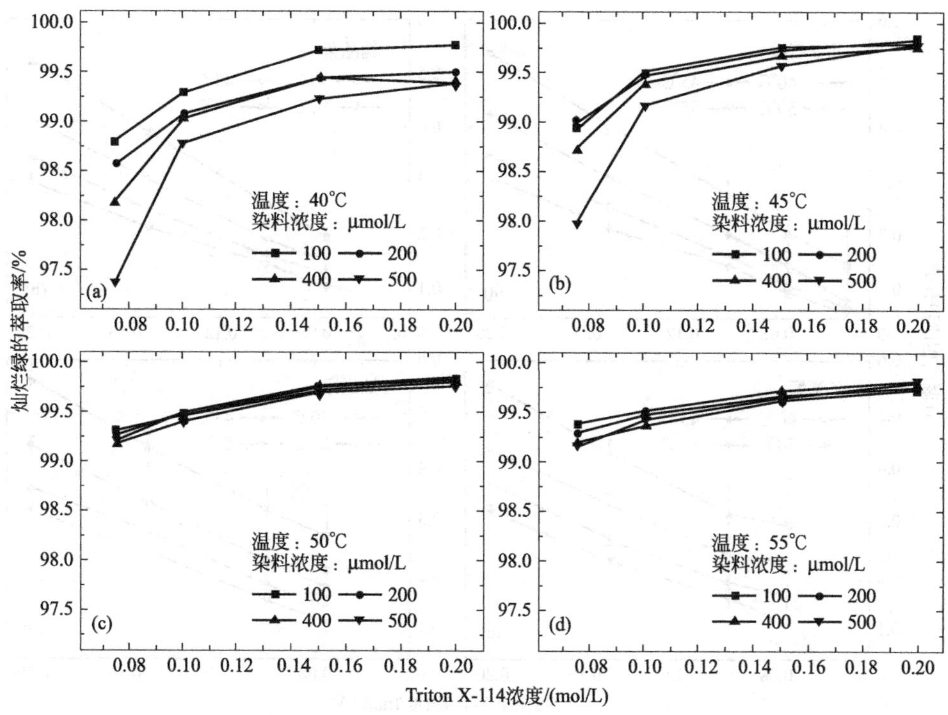

图 4-8 不同温度 Triton X-114 和灿烂绿浓度对灿烂绿萃取率的影响

义上的两相体系，因此染料的萃取率不予考虑分析。造成这种状态的原因还有待于进一步研究。

4.3.3 凝聚层相体积的改变

如图 4-9 所示，在不同 Triton X-114 浓度和操作温度下，浊点系统的凝聚层相体积分数发生改变。在一定的 Triton X-114 浓度、温度和染料类型下，凝聚层相的体积受染料浓度影响较小。在图 4-9 中，每一点代表不同染料浓度下，凝聚层相的平均体积分数，误差限代表标准偏差。在 0.15mol/L Triton X-114 和 40℃下，凝聚层相的体积分数 0.38、0.40、0.41 和 0.42 分别对应于 50μmol/L、100μmol/L、200μmol/L 和 400μmol/L 的孔雀石绿，如图 4-9(a) 所示，其平均值为 0.40，标准偏差为 0.016。图 4-9 显示出凝聚层相的体积分数随 Triton X-114的浓度增大而增大。正如前文所提到的，随着非离子表面活性剂浓度的增大胶束量升高，同时由于稀相中表面活性剂浓度基本稳定（在 CMC 左右），两种因素的结合导致凝聚层相的体积增加。操作温度可显著影响浊点系统凝聚层相

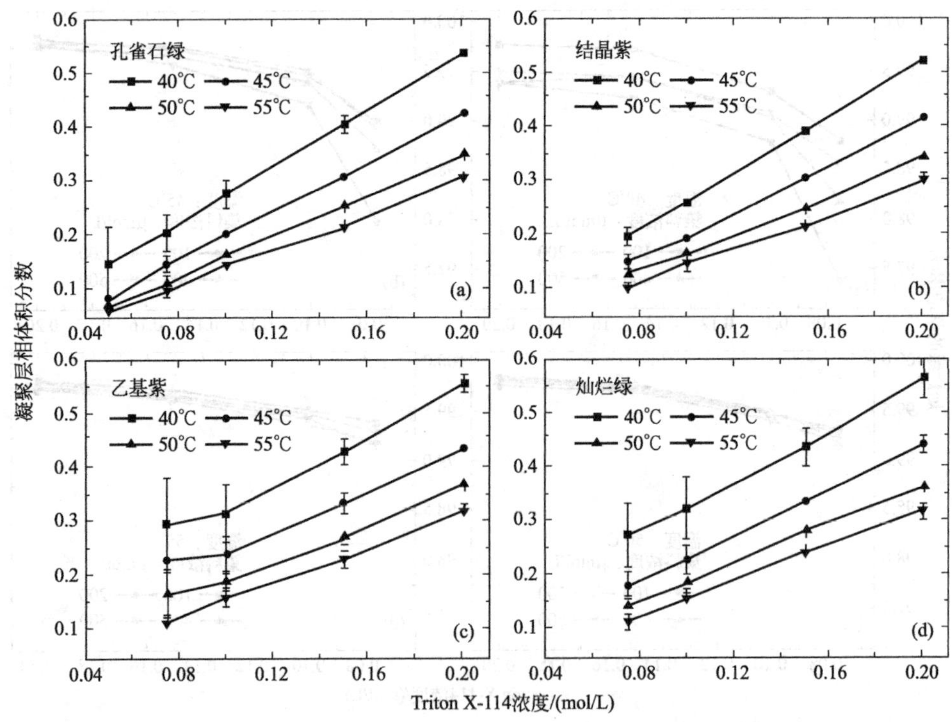

图 4-9 不同操作条件下凝聚层相体积分数的变化

的体积[11,72,81]。如前所述，随着温度的升高，非离子表面活性剂的疏水性增强。因此非离子表面活性剂分子在稀相中溶解度下降，胶束聚集加强并进一步失水，这些现象导致凝聚层相体积随温度升高而降低。

4.3.4 电解质的影响

通常情况下，在电解质的存在下，如 NaCl、$CaCl_2$、$MgCl_2$ 和 $Al(NO_3)_3$ 等，非离子表面活性剂的浊点降低[25]。因此，添加盐可有效提高浊点系统的萃取率。如图 4-10 所示，在不同的盐浓度下，染料的萃取率也不尽相同。染料萃取率随盐浓度增大而增大。例如，当 NaCl 的浓度从 0.05mol/L 增加到 0.5mol/L 时，孔雀石绿的萃取率从 97.4%增加到 98.4%。NaCl 对结晶紫、乙基紫和灿烂绿的影响具有相似的趋势。二价盐 $CaCl_2$ 对萃取率的影响也比较相似。在电解质存在情况下，非离子表面活性剂的 CMC 会降低，这一过程叫作“盐析”［如 NaCl、$CaCl_2$、$MgCl_2$ 和 $Al(NO_3)_3$ 等］[82]。因此，稀相中胶束浓度因 NaCl 和 $CaCl_2$ 的添加而降低，胶束聚集且凝聚层相由于脱水变得更加疏水。相应地，染料在凝聚

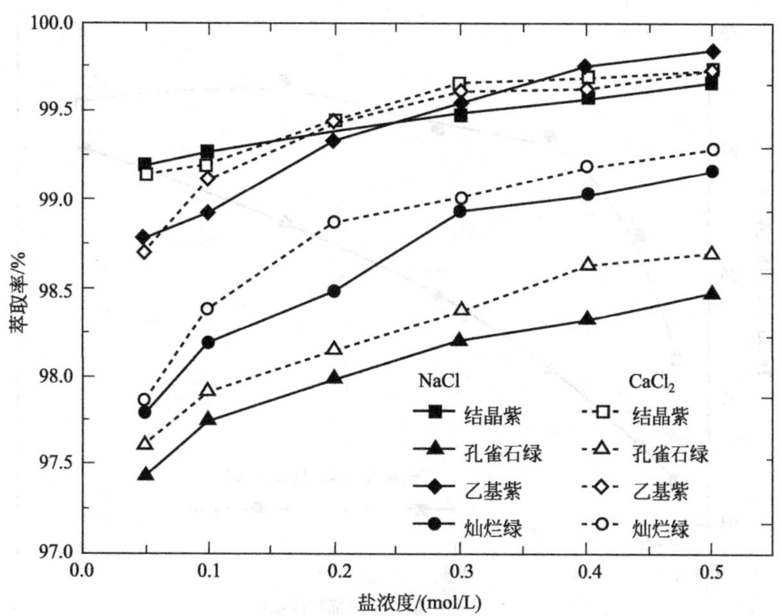

图 4-10　电解质对染料萃取率的影响

操作条件：温度为 40℃，Triton X-114 浓度为 0.075mol/L，

染料浓度为 200μmol/L

层相中的增溶进一步被强化。对于孔雀石绿和灿烂绿，在一定的盐浓度下，二价盐 $CaCl_2$ 比一价盐 NaCl 对染料的作用效果更佳。然而，对于结晶紫和乙基紫，在一定的盐浓度下，NaCl 和 $CaCl_2$ 对两种染料的作用效果没有明显差别。

4.3.5　稀相的溶剂萃取

三苯基甲烷染料废水经浊点萃取以后，稀相中少量的 Triton X-114 需要去除以达到水相无污染排放的目的。在前面的实验中，稀相中 Triton X-114 的浓度在 0.2～1.9mmol/L 之间变化。因此，人工配置 0.19mmol/L 和 1.86mmol/L 的 Triton X-114 水溶液来模拟稀相中非离子表面活性剂的最低和最高浓度。本实验选用二氯甲烷作为 Triton X-114 回收的萃取溶剂。如图 4-11 所示，随着二氯甲烷体积的增大，Triton X-114 的回收率也在升高。例如，通过把二氯甲烷和稀相的体积比由 0.052 提高到 1.0，Triton X-114（1.86mmol/L）的回收率也由 94.3%增加到 97%。对于更低浓度的 Triton X-114 水溶液（0.19mmol/L），达到相同的回收率只需要更少量的二氯甲烷。当二氯甲烷和稀相的体积比为 0.25

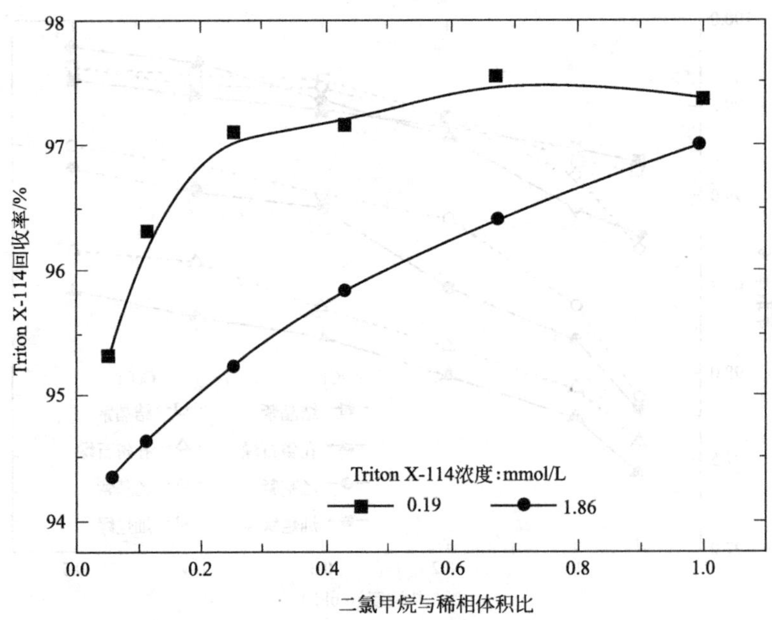

图 4-11 二氯甲烷萃取回收稀相中 Triton X-114

时，Triton X-114 的回收率已经达到 97%。进一步提高体积比到 1.0，对回收率的影响不大。在固定的二氯甲烷与稀相的体积比下，Triton X-114 的回收率随 Triton X-114 浓度的升高而降低。因为在固定的温度下[1,82]，物质在水和有机溶剂两种溶液中的分配是恒定的。因此，在固定的二氯甲烷与稀相的体积比下，降低 Triton X-114 浓度可提高其回收率。

4.3.6 增溶平衡计算

为确定在不同温度下 Triton X-114 浊点系统对几种三苯基甲烷染料的增溶能力，根据实验数据计算了系统的增溶等温线（solubilization isotherm）。增溶平衡数据是设计浊点系统的基本要求。

如图 4-12 所示，实验拟合了结晶紫、孔雀石绿、乙基紫和灿烂绿在浊点系统中不同温度下的增溶等温线。其基本规律是，随着温度的升高，每摩尔 Triton X-114 增溶的染料摩尔数增加。相同的结果在图 4-5、图 4-6、图 4-7 和图 4-8 中都有所体现。

Langmuir 吸附等温线成功地用来描述很多吸附过程[73]。在本实验中，此模

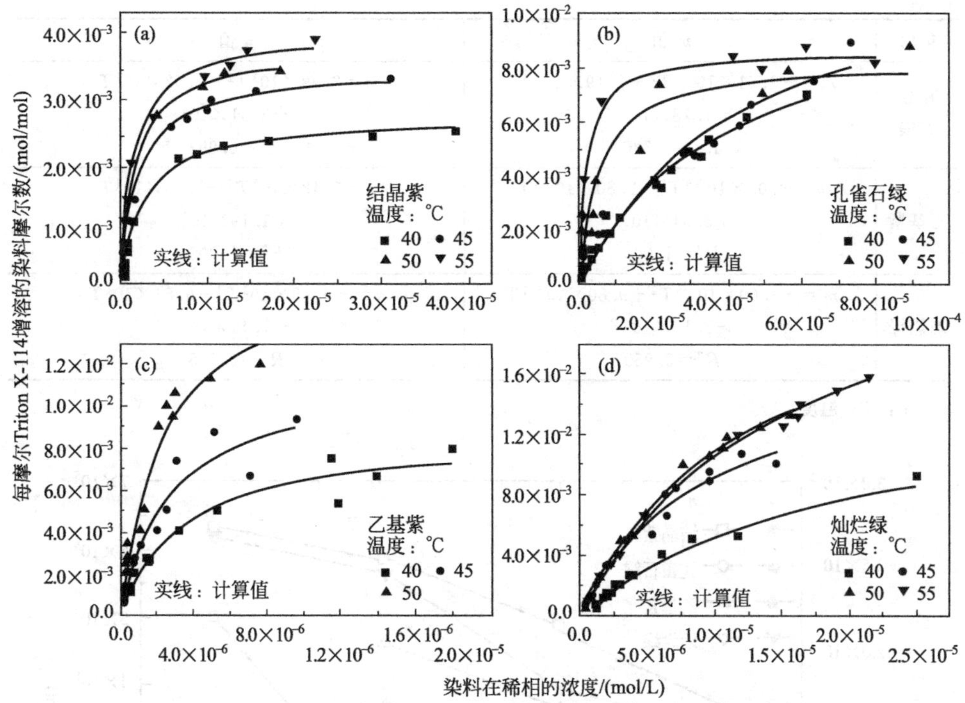

图 4-12 不同温度下四种三苯基甲烷染料在浊点系统中的增溶等温线

型用于解释四种三苯基甲烷染料在 Triton X-114 中的增溶。Langmuir 吸附方程如式(4-5) 所示：

$$q_e = \frac{mnC_e}{1+nC_e} \tag{4-5}$$

式中，q_e代表每摩尔非离子表面活性剂增溶的染料摩尔数；C_e是稀相平衡染料浓度；m（mol/mol）和 n（L/mol）是 Langmuir 常数，分别代表吸附量（solubilization capacity）和吸附能量（energy of solubilization）。通过回归分析实验数据评估每个操作温度下的 m 和 n 值，m 和 n 随温度的变化（图 4-13）符合二次方程。m 和 n 值的计算方程在表 4-5 中给出。通过式(4-5) 与 m 和 n 值的方程可以估算出去除一定量染料所需要的非离子表面活性剂摩尔数。

表 4-5 m 和 n 值的计算方程

染料	m 值	n 值
结晶紫	$m=-4.6\times10^{-6}T^2+5.22\times10^{-4}T-1.08\times10^{-2}$ $R^2=0.993$	$n=-8.33\times10^2T^2+9.77\times10^4T-2.22\times10^6$ $R^2=0.999$

续表

染料	m 值	n 值
孔雀石绿	$m=-1.51\times10^{-5}T^2+1.19\times10^{-3}T-1.18\times10^{-2}$ $R^2=0.536$	$n=2.99\times10^3T^2-2.52\times10^5T+5.31\times10^6$ $R^2=0.998$
乙基紫	$m=4.02\times10^{-5}T^2-2.80\times10^{-3}T+5.64\times10^{-2}$ $R^2=1.000$	$n=7.48\times10^2T^2-5.20\times10^4T+1.19\times10^6$ $R^2=1.000$
灿烂绿	$m=-1.60\times10^{-6}T^2+9.80\times10^{-4}T-2.17\times10^{-2}$ $R^2=0.959$	$n=-6.95\times10^2T^2+6.64\times10^4T-1.49\times10^6$ $R^2=0.735$

注：T—温度，℃。

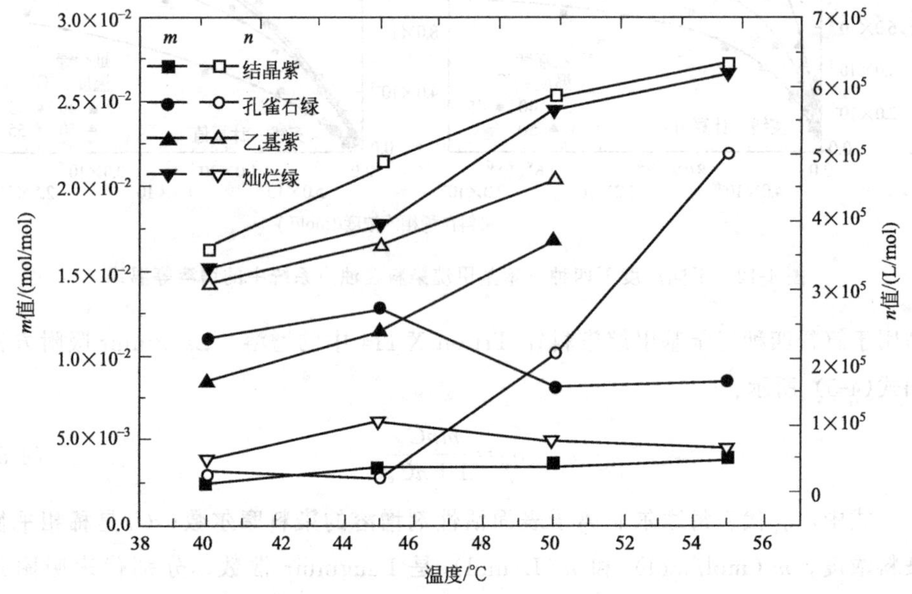

图 4-13　m 和 n 值随温度的变化

4.3.7　三苯基甲烷染料在浊点系统中的分配规律

有机物在辛醇-水两相体系中的对数分配系数 lg P，代表了有机物的疏水性。Wang 等根据有机物的疏水性建立了浊点萃取有机物的数学模型[26]。发现有机物疏水性和其在浊点系统中的增溶平衡常数具有线性关系。根据有机物的 lg P 值，可有效预测浊点系统对有机物的萃取率[26]。

如图 4-14 所示，本实验中不同温度下乙基紫的最大吸附量显著高于结晶紫的最大吸附量。根据有机溶剂在浊点系统中分配的 lg *P* 准则[26]，这主要是由于乙基紫的 lg *P* 值（3.92）高于结晶紫的 lg *P* 值（1.77）。对于灿烂绿和孔雀石绿，灿烂绿的 lg *P* 值（3.13）高于孔雀石绿的 lg *P* 值（1.69），因此在同一温度下灿烂绿的最大吸附量也高于孔雀石绿。

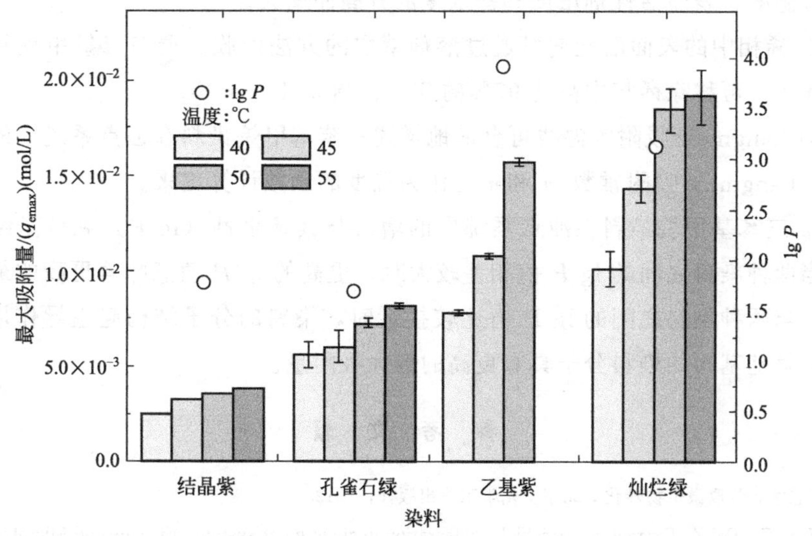

图 4-14　染料 lg *P* 与其最大吸附量（q_{emax}）之间的关系

然而，除 lg *P* 准则外[26]，在浊点系统的三苯基甲烷染料萃取实验中，发现了另外一种规律性的现象。据研究，溶质在胶束中的增溶位点和其本身的分子结构有关[83]。结晶紫、孔雀石绿、乙基紫和灿烂绿的结构与性质如表 4-4 所示。例如，结晶紫与孔雀石绿的区别是，结晶紫为三氨基类三苯基甲烷染料，而孔雀石绿为二氨基类三苯基甲烷染料。即，孔雀石绿比结晶紫少了一个二甲氨基[$—N(CH_3)_2$]。因此，尽管孔雀石绿的 lg *P* 值（1.69）小于结晶紫的 lg *P* 值（1.77），但同样温度下孔雀石绿的 q_{emax} 依然高于结晶紫的 q_{emax}。这可能和孔雀石绿缺少一个二甲氨基后空间位阻更小有关。同样地，对比乙基紫和灿烂绿，灿烂绿的 lg *P* 值（3.13）比乙基紫（3.92）更低，但由于其缺少一个二乙氨基[$—N(CH_2CH_3)_2$]，空间位阻更小，其最大吸附量依然高于后者。结果显示，三苯基甲烷染料在浊点系统中的增溶不仅受其疏水性（lg *P*）影响还与其分子结构有关。

由以上分析可知：

① 三苯基甲烷染料可显著影响 Triton X-114 水溶液的浊点。200μmol/L 的

灿烂绿、400μmol/L 的结晶紫、400μmol/L 的乙基紫和 400μmol/L 的孔雀石绿，分别可以将 Triton X-114 水溶液的浊点从 25℃ 提高到 93℃，93℃，95℃ 和 51℃。

② 使用非离子表面活性剂 Triton X-114 水溶液形成的浊点系统，可有效去除四种三苯基甲烷染料（结晶紫、乙基紫、孔雀石绿和灿烂绿）。染料的萃取率随着着温度、表面活性剂浓度和盐浓度的升高而增大。

③ 稀相中的表面活性剂可通过溶剂萃取的方法回收。调节二氯甲烷和稀相的体积比，可移除稀相中高达 97%的 Triton X-114。

④ Langmuir 吸附等温线可合适地描述三苯基甲烷染料在浊点系统中的增溶过程。Langmuir 吸附常数 m 和 n 可作为温度的函数计算出来。

⑤ 三苯基甲烷染料在浊点系统中的增溶与其疏水性（lg P）和分子结构相关。当两种染料之间的 lg P 值相差较大时，更高的 lg P 值意味着更高的最大吸附量。当两种染料之间的 lg P 值比较接近时，染料的分子结构起主导作用。缺少一个氨基基团的染料分子具有更高的最大吸附量。

参 考 文 献

［1］ 王志龙．萃取微生物转化．北京：化学工业出版社，2012.

［2］ Wang Z，Dai Z. Extractive microbial fermentation in cloud point system. Enzyme and Microbial Technology，2010，46（6）：407-418.

［3］ Miozzari G F，Niederberger P，Huetter R. Permeabilization of microorganisms by Triton X-100. Analytical Biochemistry，1978，90（1）：220-233.

［4］ Zhao F，Yu J. L-asparaginase release from *Escherichia coli* cells with K_2HPO_4 and Triton X-100. Biotechnology Progress，2001，17（3）：490-494.

［5］ Koshy L，Saiyad A H，Rakshit A K. The effects of various foreign substances on the cloud point of Triton X 100 and Triton X 114. Colloid And Polymer Science，1996，274（6）：582-587.

［6］ Bordier C. Phase separation of integral membrane proteins in Triton X-114 solution. Journal of Biological Chemistry，1981，256（4）：1604-1607.

［7］ Galera-Gómez P A，Gu T. Cloud point of mixtures of polypropylene glycol and triton X-100 in aqueous solutions. Langmuir，1996，12（10）：2602-2604.

［8］ Pourreza N，Rastegarzadeh S，Larki A. Micelle-mediated cloud point extraction and spectrophotometric determination of rhodamine B using Triton X-100. Talanta，2008，77（2）：733-736.

［9］ Laane C，Boeren S，Vos K，Veeger C. Rules for optimization of biocatalysis in organic-solvents. Biotechnology and Bioengineering，1987，30（1）：81-87.

［10］ León R，Fernandes P，Pinheiro H M，Cabral J M S. Whole-cell biocatalysis in organic media. Enzyme And Microbial Technology，1998，23（7-8）：483-500.

［11］ Quina F H，Hinze W L. Surfactant-mediated cloud point extractions：An environmentally benign alternative separation approach. Industrial and Engineering Chemistry Research, 1999, 38 (11): 4150-4168.

[12] Huddleston J G, Willauer H D, Griffin S T, Rogers R D. Aqueous polymeric solutions as environmentally benign liquid/ liquid' extraction media. Industrial and Engineering Chemistry Research, 1999, 38 (7): 2523-2539.

[13] Gil R A, Salonia J A, Gasquez J A, Olivieri A C, Olsina R, Martinez L D. Flow injection system for the on-line preconcentration of Pb by cloud point extraction coupled to USN-ICP OES. Microchemical Journal, 2010, 95 (2): 306-310.

[14] Shemirani F, Jamali M R, Kozani R R, Salavati-Niasari M. Cloud point extraction and preconcentration for the determination of Cu and Ni in natural water by flame atomic absorption spectrometry. Separation Science and Technology, 2006, 41 (13): 3065-3077.

[15] Glembin P, Kerner M, Smirnova I. Cloud point extraction of microalgae cultures. Separation and Purification Technology, 2013, 103: 21-27.

[16] Linder M B, Qiao M, Laumen F, Selber K, Hyytiä T, Nakari-Setälä T, Penttilä M E. Efficient purification of recombinant proteins using hydrophobins as tags in surfactant-based two-phase systems†. Biochemistry, 2004, 43 (37): 11873-11882.

[17] Chiang C L. Separation of cholesterol esterase in cloud point extraction system. Journal Of The Chinese Institute of Chemical Engineers, 1999, 30 (2): 171-176.

[18] Li J L, Chen B H. Solubilization of model polycyclic aromatic hydrocarbons by nonionic surfactants. Chemical Engineering Science, 2002, 57 (14): 2825-2835.

[19] An L, Deng J, Zhou L, Li H, Chen F, Wang H, Liu Y. Simultaneous spectrophotometric determination of trace amount of malachite green and crystal violet in water after cloud point extraction using partial least squares regression. Journal of Hazardous Materials, 2010, 175 (1-3): 883-888.

[20] Pourreza N, Zareian M. Determination of Orange Ⅱ in food samples after cloud point extraction using mixed micelles. Journal of Hazardous Materials, 2009, 165 (1-3): 1124-1127.

[21] Regel-Rosocka M, Szymanowski J. Direct Yellow and Methylene Blue liquid-liquid extraction with alkylene carbonates. Chemosphere, 2005, 60 (8): 1151-1156.

[22] Weschayanwiwat P, Kunanupap O, Scamehorn J F. Benzene removal from waste water using aqueous surfactant two-phase extraction with cationic and anionic surfactant mixtures. Chemosphere, 2008, 72 (7): 1043-1048.

[23] Tatara E, Materna K, Schaadt A, Bart H J, Szymanowski J. Cloud point extraction of direct yellow. Environmental Science and Technology, 2005, 39 (9): 3110-3115.

[24] Pei Y C, Wang J J, Xuan X P, Fan J, Fan M H. Factors affecting ionic liquids based removal of anionic dyes from water. Environmental Science and Technology, 2007, 41 (14): 5090-5095.

[25] Purkait M K, Banerjee S, Mewara S, DasGupta S, De S. Cloud point extraction of toxic eosin dye using Triton X-100 as nonionic surfactant. Water Research, 2005, 39 (16): 3885-3890.

[26] Wang Z. Predicting organic compound recovery efficiency of cloud point extraction with its quantitative structure-solubilization relationship. Colloids and Surfaces A: Physicochemical and Engi-

neering Aspects, 2009, 349 (1-3): 214-217.

[27] Gül Ü D, Dönmez G. Effect of surfactants on remazol blue bioremoval capacity of Rhizopus arrhizus strain growing in molasses medium. Fresenius Environmental Bulletin, 2011, 20 (10 A): 2677-2683.

[28] Baczynski T P, Pleissner D. Bioremediation of chlorinated pesticide-Contaminated soil using anaerobic sludges and surfactant addition. Journal of Environmental Science and Health—Part B Pesticides, Food Contaminants, and Agricultural Wastes, 2010, 45 (1): 82-88.

[29] Grimberg S J, Stringfellow W T, Aitken M D. Quantifying the biodegradation of phenanthrene by Pseudomonas stutzeri P16 in the presence of a nonionic surfactant. Applied and Environmental Microbiology, 1996, 62 (7): 2387-2392.

[30] Fava F, Di Gioia D. Effects of Triton X-100 and Quillaya Saponin on the ex situ bioremediation of a chronically polychlorobiphenyl-contaminated soil. Applied Microbiology and Biotechnology, 1998, 50 (5): 623-630.

[31] Mulligan C N. Environmental applications for biosurfactants. Environmental pollution (Barking, Essex: 1987), 2005, 133: 183-198.

[32] Wang Z. Bioavailability of organic compounds solubilized in nonionic surfactant micelles. Applied Microbiology and Biotechnology, 2011, 89 (3): 523-534.

[33] Saxena N, Gupta R K. Decolourization and delignification of pulp and paper mill effluent by white rot fungi. Indian Journal of Experimental Biology, 1998, 36 (10): 1049-1051.

[34] Avramova T, Stefanova L, Angelova B, Mutafov S. Bacterial decolorization of acid orange 7 in the presence of ionic and non-ionic surfactants. Zeitschrift fur Naturforschung—Section C Journal of Biosciences, 2007, 62 (1-2): 87-92.

[35] Zhang Y, Miller R M. Enhanced octadecane dispersion and biodegradation by a *Pseudomonas* rhamnolipid surfactant (biosurfactant). Applied and Environmental Microbiology, 1992, 58 (10): 3276-3282.

[36] Churchill P F, Churchill S A. Surfactant-enhanced biodegradation of solid alkanes. Journal of Environmental Science and Health—Part A Toxic/Hazardous Substances and Environmental Engineering, 1997, 32 (1): 293-306.

[37] Owsianiak M, Szulc A, Chrzanowski L, Cyplik P, Bogacki M, Olejnik-Schmidt A K, Heipieper H J. Biodegradation and surfactant-mediated biodegradation of diesel fuel by 218 microbial consortia are not correlated to cell surface hydrophobicity. Applied Microbiology and Biotechnology, 2009, 84 (3): 545-553.

[38] Liu Z F, Zeng G M, Wang J, Zhong H, Ding Y, Yuan X Z. Effects of monorhamnolipid and Tween 80 on the degradation of phenol by *Candida tropicalis*. Process Biochemistry, 2010, 45 (5): 805-809.

[39] Ding Y, Yuan X Z, Zeng G M, Liu Z F, Zhong H, Wang J. Effects of surfactants on the biodegradation of phenol by *Candida tropicalis*. Huanjing Kexue/Environmental Science, 2010, 31 (4): 1047-1052.

[40] Chan W C, You H Z. Nonionic surfactant Brij35 effects on toluene biodegradation in a composite bead biofilter. African Journal of Biotechnology, 2009, 8 (20): 5406-5414.

[41] Chan W C, You H Y. The influence of nonionic surfactant Brij 30 on biodegradation of toluene in a biofilter. African Journal of Biotechnology, 2010, 9 (36): 5914-5921.

[42] Inakollu S, Hung H C, Shreve G S. Biosurfactant enhancement of microbial degradation of various structural classes of hydrocarbon in mixed waste systems. Environmental Engineering Science, 2004, 21 (4): 463-469.

[43] Deschenes L, Lafrance P, Villeneuve J P, Samson R. The effect of an anionic surfactant on the mobilization and biodegradation of PAHs in a creosote-contaminated soil. Hydrological Sciences Journal, 1995, 40 (4): 471-484.

[44] Liu Z, Jacobson A M, Luthy R G. Biodegradation of naphthalene in aqueous nonionic surfactant systems. Applied and Environmental Microbiology, 1995, 61 (1): 145-151.

[45] Tsomides H J, Hughes J B, Thomas J M, Ward C H. Effect of surfactant addition on phenanthrene biodegradation in sediments. Environmental Toxicology and Chemistry, 1995, 14 (6): 953-959.

[46] Boonchan S, Britz M L, Stanley G A. Surfactant-enhanced biodegradation of high molecular weight polycyclic aromatic hydrocarbons by *stenotrophomonas maltophilia*. Biotechnology and Bioengineering, 1998, 59 (4): 482-494.

[47] Yang J G, Liu X, Long T, Yu G, Peng S, Zheng L. Influence of nonionic surfactant on the solubilization and biodegradation of phenanthrene. Journal of Environmental Sciences, 2003, 15 (6): 859-862.

[48] Rodriguez S, Bishop P L. Enhancing the biodegradation of polycyclic aromatic hydrocarbons: Effects of nonionic surfactant addition on biofilm function and structure. Journal of Environmental Engineering, 2008, 134 (7): 505-512.

[49] Zhou Y, Zhang J, Su E, Wei G, Ma Y, Wei D. Phenanthrene biodegradation by an indigenous *Pseudomonas* sp ZJF08 with TX100 as surfactant. Annals of Microbiology, 2008, 58 (3): 439-442.

[50] Song Y, Sun T, Xu H. Effect of surfactant TW-80 on biodegradation of PAHs in soil. Chinese Journal of Applied Ecology, 1999, 10 (2): 230-232.

[51] Yang J G, Liu X, Yu G, Long T, She P, Liu Z. Influence on the biodegradation of phenanthrene by nonionic surfactant, Tween20. Huanjing Kexue/Environmental Science, 2004, 25 (1): 53.

[52] Zang S Y, Li P J, Yang Y J, Zhang F C, Zhao H, Wang J. Study on biodegradation of BaP in the presence of surfactant TW-80. Liaoning Gongcheng Jishu Daxue Xuebao (Ziran Kexue Ban)/Journal of Liaoning Technical University (Natural Science Edition), 2007, 26 (3): 467-469.

[53] Sarma S J, Pakshirajan K. Surfactant aided biodegradation of pyrene using immobilized cells of *Mycobacterium frederiksbergense*. International Biodeterioration and Biodegradation, 2011, 65 (1): 73-77.

[54] Kim I S, Park J S, Kim K W. Enhanced biodegradation of polycyclic aromatic hydrocarbons using nonionic surfactants in soil slurry. Applied Geochemistry, 2001, 16 (11): 1419-1428.

[55] Bernardez L A. A rotating disk apparatus for assessing the biodegradation of polycyclic aromatic hy-

drocarbons transferring from a non-aqueous phase liquid to solutions of surfactant Brij 35. Bioprocess and Biosystems Engineering，2009，32 (3)：415-424.

[56] Bueno-Montes M，Springael D，Ortega-Calvo J J. Effect of a nonionic surfactant on biodegradation of slowly desorbing PAHs in contaminated soils. Environmental Science and Technology，2011，45 (7)：3019-3026.

[57] Zhu H，Aitken M D. Surfactant-enhanced desorption and biodegradation of polycyclic aromatic hydrocarbons in contaminated soil. Environmental Science and Technology，2010，44 (19)：7260-7265.

[58] Shin K H，Ahn Y，Kim K W. Toxic effect of biosurfactant addition on the biodegradation of phenanthrene. Environmental Toxicology and Chemistry，2005，24 (11)：2768-2774.

[59] Kile D E，Chiou C T. Water solubility enhancements of DDT and trichlorobenzene by some surfactants below and above the critical micelle concentration. Environmental Science and Technology，1989，23 (7)：832-838.

[60] You G，Sayles G D，Kupferle M J，Kim I S，Bishop P L. Anaerobic DDT biotransformation：enhancement by application of surfactants and low oxidation reduction potential. Chemosphere，1996，32 (11)：2269-2284.

[61] Aislabie J M，Richards N K，Boul H L. Microbial degradation of DDT and its residues a review. New Zealand Journal of Agricultural Research，1997，40 (2)：269-282.

[62] Rojas-Avelizapa N G，Rodríguez-Vázquez R，Saval-Bohorquez S，Alvarez P J J. Effect of C/N/P ratio and nonionic surfactants on polychlorinated biphenyl biodegradation. World Journal of Microbiology and Biotechnology，2000，16 (4)：319-324.

[63] Singer A C，Gilbert E S，Luepromchai E，Crowley D E. Bioremediation of polychlorinated biphenyl-contaminated soil using carvone and surfactant-grown bacteria. Applied Microbiology and Biotechnology，2000，54 (6)：838-843.

[64] Xia C，Lam J C W，Wu X，Sun L，Xie Z，Lam P K S. Levels and distribution of polybrominated diphenyl ethers (PBDEs) in marine fishes from Chinese coastal waters. Chemosphere，2011，82 (1)：18-24.

[65] Huang Y，Chen L，Peng X，Xu Z，Ye Z. PBDEs in indoor dust in South-Central China：Characteristics and implications. Chemosphere，2010，78 (2)：169-174.

[66] Deng D，Guo J，Sun G，Chen X，Qiu M，Xu M. Aerobic debromination of deca-BDE：Isolation and characterization of an indigenous isolate from a PBDE contaminated sediment. International Biodeterioration and Biodegradation，2011，65 (3)：465-469.

[67] Zhou J，Jiang W，Ding J，Zhang X，Gao S. Effect of Tween 80 and β-cyclodextrin on degradation of decabromodiphenyl ether (BDE-209) by White Rot Fungi. Chemosphere，2007，70 (2)：172-177.

[68] Ding J，Zhou J，Jiang W Y，Gao S X. Aerobic microbial degradation of polybrominated diphenyl ethers. Huanjing Kexue/Environmental Science，2008，29 (11)：3179-3184.

[69] Scott M J，Jones M N. The biodegradation of surfactants in the environment. Biochimica et Biophysica Acta—Biomembranes，2000，1508 (1-2)：235-251.

[70] Saichek R E，Reddy K R. Electrokinetically enhanced remediation of hydrophobic organic compounds

in soils: a review. Critical Reviews in Environmental Science and Technology, 2005, 35 (2): 115-192.

[71] Gregory P. Dyes and dye intermediates//Kroschwitz J I. Encyclopedia of Chemical Technology: Vol 8. New York: John Wiley and Sons, 1993: 544-545.

[72] Hinze W L, Pramauro E. A critical review of surfactant-mediated phase separations (cloud-point extractions): Theory and applications. Critical Reviews in Analytical Chemistry, 1993, 24 (2): 133-177.

[73] Purkait M K, DasGupta S, De S. Performance of TX-100 and TX-114 for the separation of chrysoidine dye using cloud point extraction. Journal of Hazardous Materials, 2006, 137 (2): 827-835.

[74] Pourreza N, Elhami S. Removal of malachite green from water samples by cloud point extraction using Triton X-100 as non-ionic surfactant. Environmental Chemistry Letters, 2008, 8 (1): 53-57.

[75] Wang Z. Extractive whole cell biotransformation in cloud point system//Wendt P L, Hoysted D S. Non-Ionic Surfactants. New York: Nova Science Publishers, 2010: 83-136.

[76] López-Montilla J C, Pandey S, Shah D O, Crisalle O D. Removal of non-ionic organic pollutants from water via liquid-liquid extraction. Water Research, 2005, 39 (9): 1907-1913.

[77] Liang R, Wang Z, Xu J H, Li W, Qi H. Novel polyethylene glycol induced cloud point system for extraction and back-extraction of organic compounds. Separation and Purification Technology, 2009, 66 (2): 248-256.

[78] Wang Z, Liang R, Xu J H, Liu Y, Qi H. A closed concept of extractive whole cell microbial transformation of benzaldehyde into L-phenylacetylcarbinol by saccharomyces cerevisiae in novel polyethylene-glycol-induced cloud-point system. Applied Biochemistry and Biotechnology, 2010, 160 (6): 1865-1877.

[79] Pan T, Wang Z, Xu J H, Wu Z, Qi H. Extractive fermentation in cloud point system for lipase production by *Serratia marcescens* ECU1010. Applied Microbiology and Biotechnology, 2010, 85 (6): 1789-1796.

[80] Pan T, Wang Z, Xu J H, Wu Z, Qi H. Stripping of nonionic surfactants from the coacervate phase of cloud point system for lipase separation by Winsor II microemulsion extraction with the direct addition of alcohols. Process Biochemistry, 2010, 45 (5): 771-776.

[81] Wang Z, Xu J H, Zhang W, Zhuang B, Qi H. Cloud point of nonionic surfactant Triton X-45 in aqueous solution. Colloids and Surfaces B: Biointerfaces, 2008, 61 (1): 118-122.

[82] Mukherjee P, Padhan S K, Dash S, Patel S, Mishra B K. Clouding behaviour in surfactant systems. Advances in Colloid and Interface Science, 2011, 162 (1-2): 59-79.

[83] Myers D. Surfaces, interfaces, and colloids: principles and applications, 2nd edition. New York: John Wiley and Sons, 1999.

第 5 章
浊点系统中三苯基甲烷染料的微生物脱色

5.1 浊点系统中的生物过程

5.1.1 溶质的增溶

增溶作用被定义为“通过引入额外的两亲性化合物或组分，制备通常不溶于或微溶于给定溶剂的热力学稳定的各向同性溶液”[1]。溶质在表面活性剂胶束中的增溶机理已经被广泛研究，如烃类增溶在胶束疏水核心、极性更强的物质增溶在栅栏区域、极性物质增溶在水合层以及直接溶解在双电层的溶质等[2]。不仅疏水性溶质，有些亲水性溶质也可以溶解到表面活性剂胶束中。极性化合物，特别是那些带有氢键官能团或带有 C—H…π 键芳烃的化合物，也发现了增溶作用[3]。非离子表面活性剂水溶液的浊点系统成为了一种重要的分离介质，而浊点萃取已被广泛应用。

浊点萃取作为一种环保的方法在分离领域中的应用由来已久，并且受到普遍关注[4]。然而，对浊点体系凝聚相增溶的研究相对较少。其主要障碍是难以确定自由溶质浓度，即难以区分表面活性剂超分子组装体和水溶液中的溶质。水溶液中非离子表面活性剂的超分子组装体的不同类型导致其增溶能力的差异。低浓度苯酚的增溶实验表明，溶质在稀相和凝聚层相之间的浓度差很小[5]。然而，随着苯酚浓度的增大，凝聚层相中的苯酚浓度显著增大[6]。凝聚层相中超分子组装体的结构通过添加苯酚来改变，以降低非离子表面活性剂水溶液的浊点，这类似于超分子组装体随着温度的升高而从层状结构变为囊状结构[1]。增溶能力的不同可能与凝聚相超分子组装结构的差异有关。在非离子表面活性剂 Brij 35 的浊点体系中也报道了类似的现象，1mol 表面活性剂溶解的苯酚从稀相中的 2.9mol 增加到凝聚相中的 7.5mol[7]。浊点系统在稀相和凝聚层相中的不同增溶能力进一步证明，增溶与非离子表面活性剂的超分子组装结构有关。

5.1.2 浊点系统的生物相容性

非水介质的生物相容性是非水介质应用于全细胞微生物转化或降解的前提条

件。一些表面活性剂浓度相对较低的非离子表面活性剂胶束溶液被用于提高疏水化合物的生物利用度。然而，非离子表面活性剂对微生物细胞的渗透性使其对微生物有毒。

利用分枝杆菌筛选的多种非离子表面活性剂水溶液中，只有浊点低于培养温度且相分离形成稀相和凝聚相的 Triton X-114 具有生物相容性。分散在水中的 Brij 30 与 Triton X-45 以及浊点高于培养温度的表面活性剂，对微生物有毒[8,9]。进一步研究了非离子表面活性剂水溶液的相分离与生物相容性的关系。相同系列的非离子表面活性剂 Triton X-100、Tritonx-114 和 Tritonx-45 具有相同的疏水部分，但环氧乙烷单元数目不同。Triton X-114 具有生物相容性，但 Triton X-100 对微生物分枝杆菌具有毒性。尽管任意比例的 Triton X-114 与 Triton X-100 的混合物都能溶解疏水性基质，但仅当 Triton X-114 与 Triton X-100 高于一定体积比时，溶液的浊点低于培养温度，系统具有生物相容性[8]。相分离仅在 Triton X-45 与 Triton 114 的体积比低于 40%时发生。在这个比例以上，混合非离子表面活性剂水溶液的浊点均高于微生物培养温度。

研究了 PEG 诱导的浊点系统与水-有机溶剂两相分配系统的生物相容性，累积的生物量和残留的葡萄糖作为量化指标。在有机溶剂介导的水-有机溶剂两相分配系统中，生物相容性随 lg P 的增大而增强（lg P 被定义为辛醇-水分配系数，通常用于指示有机溶剂的极性[10]）。根据 E_t数据，非离子表面活性剂的浊点系统比水-有机溶剂两相分配系统的极性更强[11]。然而，PEG 诱导的浊点系统中酿酒酵母保持了生物相容性，这说明该体系具有相对较高极性的生物相容性环境。

有毒的非离子表面活性剂在相分离条件下对微生物变得生物相容。浊点系统的生物相容性原理与水-有机溶剂两相分配系统相似[9,12-14]。不同之处在于浊点系统中的凝聚层相代替了水-有机溶剂两相分配系统中的有机溶剂辅助相。众所周知，大多数有机溶剂对微生物有毒，只能添加有限浓度的有机溶剂形成单相有机溶剂系统以提高某些有机基质的溶解度，该方法已应用于微生物转化[15]。有机溶剂对微生物的临界抑制浓度随着溶剂极性的增强而增大[10,13]。然而，有机溶剂在水-有机溶剂两相分配系统中对微生物的毒性是由其在水环境中的实际浓度而不是总有机溶剂浓度决定的。lg P 准则表明，lg P 大于 4 的有机溶剂具有生物相容性[10,16,17]。

类似地，有毒的非离子表面活性剂在高于其浊点的温度下浓缩进入凝聚层，而稀释相中的表面活性剂浓度相对较低。低浓度非离子表面活性剂胶束溶液具有生物相容性，已在生物技术中得到应用[18-20]。类似于 lg P 作为有机溶剂的参数，非离子表面活性剂的浊点可以用作非离子表面活性剂的选择标准。然而，在浊点系统中用非离子表面活性剂代替辅助相具有更多优点。例如，非离子表面活性剂的不挥发性和非固定性满足了绿色溶剂的需求，避免了在有氧条件下水-有机溶剂两相分配系统中存在的爆炸危险[21]。在浊点系统中，非离子表面活性剂相对较高的极性和生物相容性对于中等极性底物/产物的全细胞微生物转化具有应用潜力。由于不能同时实现产品的生物相容性和提取，因此水-有机溶剂两相系统无法达到此目的[22,23]。

5.1.3 浊点系统中的生物转化

传统的生物转化底物产物 lg P 值都处于 2～4 之间，例如有机酮、醇、酸及酯等，底物产物毒性较大[17]。在微生物培养过程中，受到底物抑制、产物反馈抑制、产物生物降解等因素限制，产物浓度较低[24]。两相系统的萃取微生物转化可有效解决这一难题[10]。用于萃取微生物转化的两相系统主要包括：水-有机溶剂两相系统[25-27]、双水相系统[28,29]、室温离子液体系统[30]和浊点系统[8,31-36]。几种两相系统在萃取微生物转化中的优缺点如表 5-1 所示。

表 5-1 几种两相系统在萃取微生物转化中的优缺点

两相系统	优点	缺点
水-有机溶剂两相系统	研究基础好，应用广泛，容易回收	易爆炸，污染环境，生物毒性大
双水相系统	生物相容性好，绿色溶剂	溶剂昂贵，浓度高时黏度大
室温离子液体系统	极性谱宽、溶剂选择性大，不挥发、绿色溶剂	基础理化参数不足
浊点系统	生物相容性好，绿色溶剂，操作容易	溶剂回收困难

上海交通大学的王志龙研究员团队在浊点系统中的萃取微生物转化方面的研究较为深入[37]。王志龙等在浊点系统中成功实现甾簇医药品的重要中间体雄甾-1,4-二烯-3,17-二酮（ADD）的萃取微生物转化，此萃取微生物转化技术克服了底物/产物的溶解性差、底物/产物对微生物产生毒性、产物被微生物进一步降解等传统微生物转化技术的缺点，将 ADD 的产量由原来的不足 0.5g/L 提高到 10g/L，满足了工业生产需求[31]。该团队又陆续开发了 6-APA[35,38] 及 L-PAC[39] 等在浊点系统中的萃取微生物转化技术，皆取得可喜的成绩。

(1) 甾醇的微生物转化

浊点系统中甾醇的微生物转化产生类固醇中间体 ADD，是非水生物转化的典型例子。其过程涉及底物的低溶解度、底物和产物的抑制以及微生物对产物的进一步降解[40-43]。在水-有机溶剂两相系统和双水相系统中都开展过全细胞微生物转化[40,41]。然而，生物过程受限于有机溶剂的生物相容性或双水相体系的增溶作用。分枝杆菌在 Triton X-114 或 Triton X-114 和 Triton X-100 混合表面活性剂浊点系统中保持其生物相容性。浊点系统提高了底物溶解度，进而提高了甾醇的生物利用度[8,31]。静息细胞微生物转化进一步确认浊点系统消除了底物和产物的抑制，并防止产物进一步降解[32]。所有这些都使得浊点系统中的生物催化活性从 1mg/d 提高了 24 倍[32]。通过使用生长细胞的微生物转化法，将产品的浓度提高到 10g/L 以上[8]；通过在浊点系统中应用静息细胞微生物转化法，产品浓度进一步提高到 12g/L[34]。

(2) 手性 1-苯乙醇的不对称生物合成

酿酒酵母将苯乙酮微生物转化为手性 1-苯乙醇可能是最流行的模型反应之一，它涉及辅因子 NAD(P)H 与活细胞的再生[44-47]。然而，底物/产物对微生物的毒性使得全细胞微生物的利用[48]，即使采用固定细胞技术[44-46]，也只有在底物浓度非常低的条件下才能实现自我再生和自我增殖[47]。在水溶液中，酿酒酵母可以生长的有毒底物和产物浓度分别为 0.3%和 0.4%（体积分数）。在浊点系统中，这一限值已分别提高到 0.6%和 0.7%（体积分数）。微生物全细胞作为生物催化剂的再利用表明其生物催化活性至少保持了 4 天[49]。

(3) 微生物转化生产 L-苯乙酰甲醇

啤酒酵母的丙酮酸脱羧酶通过丙酮酸的非氧化脱羧，然后羰基化为苯甲醛，可以实现苯甲醛向 L-苯乙酰甲醇（L-PAC）（用于生产 L-麻黄碱和相关伪麻黄碱的中间体）的生物转化。为了使该过程经济可行，应通过细胞代谢葡萄糖产生丙酮酸[50,51]。全细胞微生物转化被认为可用于这一过程。但 lg P 值为 0.6 的产物只能用极性相对较高的有机溶剂（如戊醇、辛醇等）进行有效萃取[52,53]。而细胞很难在相对较高极性的水-有机溶剂两相系统中维持生物相容性[23]。即使是固定化细胞技术可以在固定化基质内建立毒性底物和/或产物梯度，并且毒性底物/产物由于扩散限制而浓度减小[54]，对这种微生物转化也不是非常有效[55]。然而，酿酒酵母在浊点系统或聚乙二醇诱导的浊点系统中保持着生物相容性，尽管其极性相对较高。同时，在微生物转化过程中，具有相对较高极性的产物 L-PAC 也被萃取到凝聚层相中[39]。在 PEG 诱导的浊点系统中萃取相对较高极性的产物，将填补水-极性有机溶剂两相分配系统用于中等极性底物/产物的微生物萃取转化所留下的产物极性谱的空白[22,23]。同时，通过加入 PEG 降低水溶液中非离子表面活性剂的浊点，也扩大了生物相容性非离子表面活性剂的选择范围。在常规微生物转化温度下，全细胞微生物转化的浊点相对较高。在PEG诱导的浊点系统中，微生物转化的进一步优化导致产物浓度提高到约8g/L[56]。

5.1.4 浊点系统中的萃取发酵

微生物通常生长在温和条件下（中等温度、压力、pH值、盐浓度）的水溶液中，然而能在高（低）温、高（低）压、高（低）盐浓度、高（低）pH值等极端条件下生长的极端微生物相继发现。1989年，日本学者Inoue等发现假单胞菌（*Pseudomonas*）能在高浓度甲苯中存活[57]。这一有机溶剂耐受性微生物的发现，为非水溶剂中微生物发酵的研究带来了崭新的开端。寻找既有有机溶剂耐受性又具有优良催化活性的生物催化剂成为重要的研究方向[58]。

有机溶剂对微生物的毒性主要在于破坏细胞膜的结构，微生物对有机溶剂的耐受性则在于细胞膜对环境条件的适应性[59]。研究表明有机溶剂对微生物的毒性取决于有机溶剂在细胞膜中的积累浓度。细胞膜中有机化合物的积累浓度与有机溶剂的极性成反比，即非极性有机溶剂具有生物毒性[60]。经验规则表明强非极性有机溶剂对微生物毒性较小，而中等非极性有机溶剂对微生物毒性较强[14]。这可归因于强非极性溶剂在水中具有很小的溶解度。中等极性有机物的微生物转化一直面临着较大的困难，相继出现了双水相系统[61]、室温离子液体系统[62]中等极性产物的萃取微生物发酵技术。

浊点系统最初用于金属离子的萃取浓缩[63]，后来用于膜蛋白的分离[64]。本团队进一步利用浊点系统进行萃取微生物发酵[8]。在研究了小分子有机化合物的萃取微生物发酵的基础上，选择极性相对较强的脂肪酶为模型，探索浊点系统中生物大分子的萃取发酵及其下游分离。脂肪酶属于诱导酶，因此在发酵液中添加三酰基甘油、脂肪酸、水解酯类、吐温、胆汁盐、甘油等碳源作为诱导剂有利于产酶[65,66]。

生物产业已经成为当今世界各国和地区重点发展的高新技术产业，而与人们日常生活联系最为密切的则是传统的微生物发酵产业。为了提高发酵过程的生产水平，人们首先考虑的是菌种选育或构建基因工程菌，往往忽略了生物反应器工程问题中必须考虑的工艺变化和过程优化。一些生物反应过程，如生物转化、生物降解等，在常规介质中常面临底物在水溶液中溶解度小、底物/产物抑制微生物生长、产物降解等实际困难[67]。将发酵过程和产物萃取过程整合为一体的萃取发酵技术解决了上述难题[15]。

萃取发酵，顾名思义，就是一边发酵一边萃取，将产物的发酵生产过程和萃取分离过程合二为一。其基本思想是利用分离科学领域里的介质工程原理，指导发酵生产。介质工程是指向水溶液中加入具有生物相容性和本身不被生物降解的组分，以达到底物缓释、产物原位去除的目的[68]。其中，两相分配系统的应用使传统发酵系统的各种问题得以解决[15]。

两相分配系统常用于分离科学领域，包括有机溶剂萃取、双水相萃取、反胶束萃取、超临界流体萃取和离子液体萃取等。萃取发酵是将上述两相分配系统作为传统发酵过程的反应介质，从而避免了底物抑制、产物抑制以及底物溶解度小等问题。

(1) 脂肪酶的萃取微生物发酵

采用黏质沙雷菌 ECU1010（*Serratia marcescens* ECU1010）作为发酵菌种，完成了浊点系统中脂肪酶萃取发酵的研究[69]，该研究也属大分子在浊点系统中萃取发酵的首次报道。黏质沙雷菌 ECU1010 脂肪酶是合成心血管病药物地尔硫䓬前体——(±)-MPGM［*trans*-3-（4′-甲氧基苯基）缩水甘油酸甲酯］的重要催化剂[70,71]。与萃取微生物转化不同的是，萃取发酵无需添加前体物质，微生物细胞直接利用培养基营养物质进行产物的合成。

(2) 红曲色素的渗透萃取发酵

传统生物技术为提高红曲色素的产量，通常在液体深层培养基中添加树脂等色素的吸附剂，以达到萃取发酵的目的。但菌体细胞内部色素含量仍然很高，胞外分泌速率慢导致了产物反馈抑制，影响了色素的进一步合成。Hu 等利用 Triton X-100 胶束溶液实现了红曲色素的胞内渗透萃取发酵[72]。非离子表面活性剂形成的胶束根据相似相容原理，对菌体细胞膜造成一定物理损伤，增强了通透性，促进了胞内色素的胞外分泌[73]。同时，表面活性剂进一步影响了细胞的代谢。细胞膜的不饱和脂肪酸含量的增加增强了细胞膜流动性。胞内色素向胞外的快速转移使得胞内色素含量降低，底物反馈抑制作用解除，色素产量随之提高。另外，表面活性剂的类型、浓度、添加时间除影响色素产量外，也进一步影响了色素的组成[74]。通过调整萃取发酵系统组成，可将红曲色素从胞内转移到胶束中，一方面避免了产物的胞内生物降解，另一方面也提高了红曲色素的产量[75]。

5.1.5 浊点系统中的生物降解

浊点系统中多环芳烃的生物降解已被证实有效，但目前研究尚处于初级阶段，受关注较少[76]。这主要是因为表面活性剂胶束水溶液的浊点对添加物的改变非常敏感[77]。在有机物生物转化过程中，随着培养基中基质和代谢产物浓度的变化，浊点系统存在体系崩溃的风险[78]。在 Brij 30（质量分数为 5%）形成的浊点系统中，菲表观溶解度的增大强化了恶臭假单胞菌 DSMZ8368 对菲的降解[79]。我们团队也开展了浊点系统中多环芳烃的生物降解[80]，发现不同表面活性剂形成的浊点系统，其对多环芳烃降解的影响也是不同的。与 Triton X-114＋Triton X-45、Triton X-114、Triton X-45 和 Brij 30 等形成的浊点系统相比，Brij 30＋Tergitol TMN-3 体系更有利于鞘氨醇单胞菌降解菲。

生物利用度是指有机化合物可被微生物直接降解的部分。表面活性剂可以在溶液中形成胶束，从而增大多环芳烃的表观溶解度。但是在胶束体系中，表面活性剂本身的毒性会限制微生物降解多环芳烃。有研究表明，在浊点系统中所使用的非离子表面活性剂，其在水环境中表现为低毒和可生物降解的特性[77]。与表面活性剂胶束水溶液相比，浊点系统具有良好的生物相容性和易回收性。本团队使用由混合型非离子表面活性剂 Brij 30 和 TMN-3 分别形成胶束体系和浊点系统，并用萘和菲进行生物毒性实验[81]。结果发现在胶束体系中多环芳烃的生物毒性会随着表面活性剂的浓度增大而增强。但在浊点系统中，多环芳烃的生物毒性一直处于较低水平。

这主要是两种系统中多环芳烃不同的生物利用度造成的。与胶束体系相比，浊点系统的凝聚相消除了底物和产物对污染物的抑制作用，从而增强了生物降解性[81]。在浊点系统中，由于浊点系统的增溶作用，多环芳烃被提取到凝聚相。因此，由多环芳烃疏水性引起的微生物降解速率限制被打破，从而促进了转移和吸收。由于大量的多环芳烃被提取到凝聚相，在稀相中只留下少量的多环芳烃。而亲水的微生物细胞存在于表面活性剂浓度很低的稀相中，因此浊点系统中多环芳烃表现出的微生物毒性更低。

5.2 萃取微生物脱色的菌种培养与产物检测

本章致力于将浊点系统应用到三苯基甲烷染料的生物脱色，实现浊点系统中染料的萃取微生物脱色。这是浊点系统在染料生物降解中的首次尝试。

前文已经提到，三苯基甲烷染料广泛应用于纺织、食品和医药工业[82]。染料分子骨架的稳定性，具有微生物及哺乳动物细胞毒性并造成环境污染[83]。物理或化学的方法已经成功用于这类染料的治理，但是其消耗高、花费大的缺点阻碍了这些方法的实际应用[84]。

生物脱色是处理三苯基甲烷染料污染的良好选择。有研究发现三苯基甲烷染料可被生物降解为二氧化碳和水[85,86]。然而，自然界中更多的微生物只能将这些化合物转化为有毒的副产物。例如，下面的这些结晶紫降解菌只能将结晶紫转化为隐性结晶紫，如：*Pseudomonas otitidis* WL-13[87]，通过吸附；*Pseudomonas putida*，通过脱甲基[88]；*Citrobacter* sp. KCTC 18061P，通过加氢[89]。有毒副产物的产生是这些菌种实际应用的主要障碍。将浊点系统用于三苯基甲烷染料的生物脱色有望克服这一难题。浊点系统是非离子表面活性剂水溶液在一定温度下分相形成的两相系统，包含凝聚层相和稀相。浊点系统已成功应用于萃取微生物转化和萃取发酵[36,69]。本章主要探索浊点系统在三苯基甲烷染料嗜水气单胞菌 DN322p 微生物脱色中的应用。

5.2.1 萃取微生物脱色的菌种及培养基

实验用的菌种嗜水气单胞菌 DN322p 由嗜水气单胞菌 DN322 复筛得来，详情已在第 3 章中介绍。

LB 培养基用于菌体培养。在 LB 培养基中加入 200mg/L 的染料形成脱色培养基。将一定量的非离子表面活性剂加入脱色培养基中用于萃取微生物脱色实验。混匀后，取 30mL 培养基转移到 250mL 三角瓶中并接种嗜水气单胞菌 DN322p。脱色培养在 30℃、200r/min 摇床中进行。

脱色培养一段时间后，培养液在 25℃、8000r/min 下离心 20min。生物量以细胞湿重（wet cell weight，WCW）表示。上清液用于分析染料浓度。

5.2.2 染料的分析方法

由于萃取微生物脱色培养基上清液中存在非离子表面活性剂，测量过程中会发生浑浊影响染料浓度测定。因此，浊点系统上清液用 75%的乙醇溶液稀释。稀释的水溶液用于分析染料浓度。染料浓度采用分光光度法测定。结晶紫、孔雀石绿、乙基紫和灿烂绿的最大吸收峰分别在 582nm、625nm、595nm 和 616nm。

染料脱色率按式(5-1) 计算：

$$染料脱色率=\left(1-\frac{上清液中的染料浓度}{染料初始浓度}\right)\times 100\% \quad (5\text{-}1)$$

萃取微生物脱色后，结晶紫和隐性结晶紫在浊点系统中的分配情况通过TLC分析。TLC条件为：采用GF254硅胶板，层析液为乙酸乙酯：水：冰醋酸＝90：9：1[90]。结晶紫和隐性结晶紫的比移值分别为0.78和0.43。

5.3 萃取微生物脱色过程与机理

5.3.1 非离子表面活性剂的筛选

实验筛选了几种非离子表面活性剂（40g/L）用于嗜水气单胞菌DN322p的萃取微生物脱色，以期获得良好生物相容性和染料脱色率的体系（图5-1）。培养24h以后，测定WCW和染料脱色率。不同非离子表面活性剂溶液条件下，

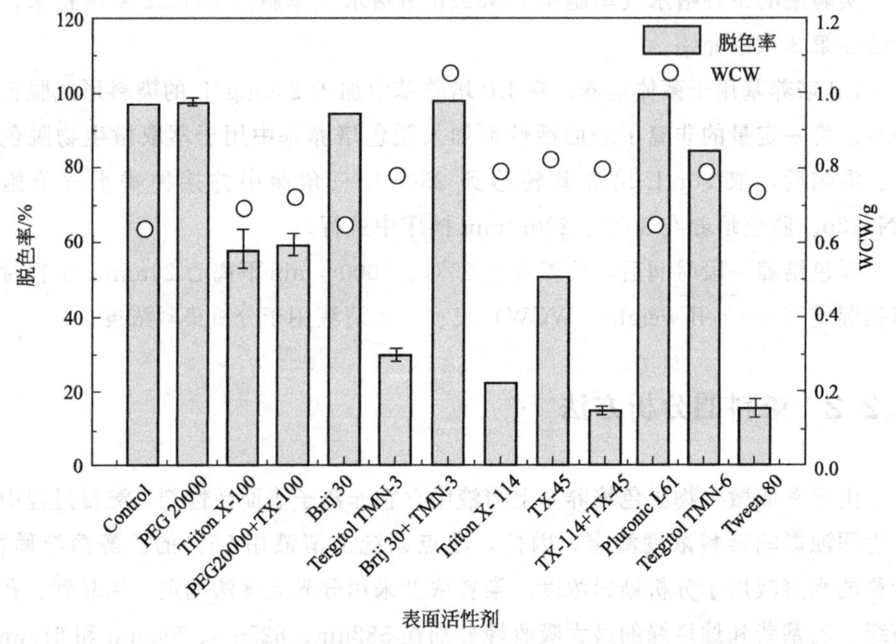

图5-1　非离子表面活性剂用于*A. hydrophila* DN322p的萃取微生物脱色

WCW 没有太大差别，说明所选用的非离子表面活性剂不影响菌株嗜水气单胞菌 DN322p 的生长繁殖，具有生物相容性。但是，在不同的非离子表面活性剂水溶液中，结晶紫的脱色率受影响较大。因为在非离子表面活性剂水溶液中，染料被萃取到胶束中，对染料的生物利用度造成影响。在 Triton X-100、PEG 20000＋Triton X-100、Tergitol TMN-6 和 Tween 80 胶束系统中，在 Triton X-45 分散系统中，以及在 Tergitol TMN-3、Triton X-114 和 Trtion X-114＋Triton X-45 系统中，结晶紫的脱色率较低。在 PEG 20000、Brij 30、Pluronic L61 和 Brij 30＋TMN-3 系统中，染料脱色率较高。PEG 20000 是单相系统，Brij 30 形成分散系统，容易导致二次污染。Pluronic L61 在离心后会在离心管底形成微液滴，这种微液滴难以与细胞沉淀分离。对于 Brij 30＋TMN-3 系统，混合的非离子表面活性剂在水溶液中室温下形成浊点系统。与对照相相比，40g/L 的 Brij 30＋TMN-3（1：1）浊点系统同时满足生物相容性及高脱色率的要求。

在传统的水溶液中，染料的微生物脱色受到染料抑制的限制。底物产物毒性及未脱毒产物是令微生物脱色方法难以在染料废水处理中应用的主要因素。浊点系统已成功应用于萃取微生物转化[36,39]和微生物发酵[69]。在染料废水处理领域，浊点系统中的萃取微生物脱色是有一次有趣的尝试。生物相容性是浊点系统用于萃取微生物脱色的前体条件。生物相容性与非离子表面活性剂的种类和浓度，以及微生物自身有关。实验发现 *Saccharomyces cerevisiae* 在 Triton X-100 的非离子表面活性剂水溶液或浊点系统中具有生物活性[91]，但在 Triton X-114 的分散体系中无法生长[36]。如图 5-1 所示，嗜水气单胞菌 DN322p 的活性不受非离子表面活性剂种类的影响，说明这些非离子表面活性剂水溶液体系都具有良好的生物相容性。然而，脱色率受影响较大。Pluronic L61 由于在离心管底形成微液滴难以与细胞沉淀分离，导致 WCW 及后续实验的分析变得困难。虽然结晶紫在几种单一（PEG 20000）或分散（Brij 30）的溶液中脱色率较高，但由于会造成二次污染给排水带来困难，因此也不适合作为脱色体系。Brij 30＋TMN-3 形成的混合非离子表面活性剂浊点系统同时具有良好的生物相容性及高脱色率，可作为染料萃取微生物脱色体系。

5.3.2 混合表面活性剂浓度对脱色率的影响

如图 5-2 所示，在不同浓度 Brij 30＋TMN-3 混合非离子表面活性剂形成的浊点系统中，染料的萃取微生物脱色效果是不同的。Brij 30 和 TMN-3 比例在 1：1 时，实验获得相对高的脱色率及稀相的低染料残留。Brij 30＋TMN-3 的总

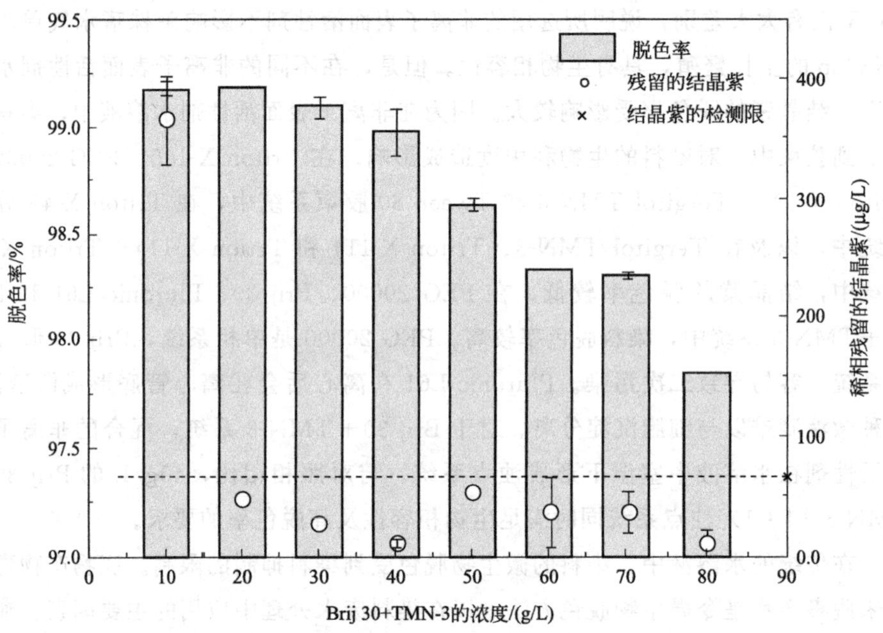

图 5-2　非离子表面活性剂浓度对结晶紫脱色率的影响

浓度在 20g/L 以内时，染料脱色率保持稳定；当高于这一浓度时，脱色率有所下降。这时，可以明显观察到有更多的结晶紫进入到浊点系统中的凝聚层相而没有被微生物脱色。当 Brij 30＋TMN-3 的浓度为 10g/L 时，染料脱色率较高，但同时稀相中染料残留也最高。综合染料脱色率和稀相染料残留，Brij 30＋TMN-3 混合非离子表面活性剂的最终浓度选定在 20g/L。此时，稀相中染料残留浓度低于分光光度法的检测限（63μg/L）。稀相中染料残留更低将有利于染料萃取微生物脱色后的污水排放。

非离子表面活性剂浓度是浊点系统相行为的重要参数[37]。一般情况下，凝聚层相的体积随表面活性剂浓度的增高而增大，因此，凝聚层相的萃取容量也随之增大。然而，在本实验中，高非离子表面活性剂浓度下，染料脱色率较低主要是因为更多的结晶紫分配进入凝聚层相当中导致其生物利用度下降。另外，稀相中染料残留越低越有利于水相的脱毒。最终，Brij 30＋TMN-3 的总浓度选在 20g/L，既有利于染料脱色又保持了水相中染料的低浓度。

5.3.3　浊点系统中的细胞生长曲线与染料脱色曲线

实验对比了对照系统和浊点系统中细胞生长及染料脱色曲线，如图 5-3 所

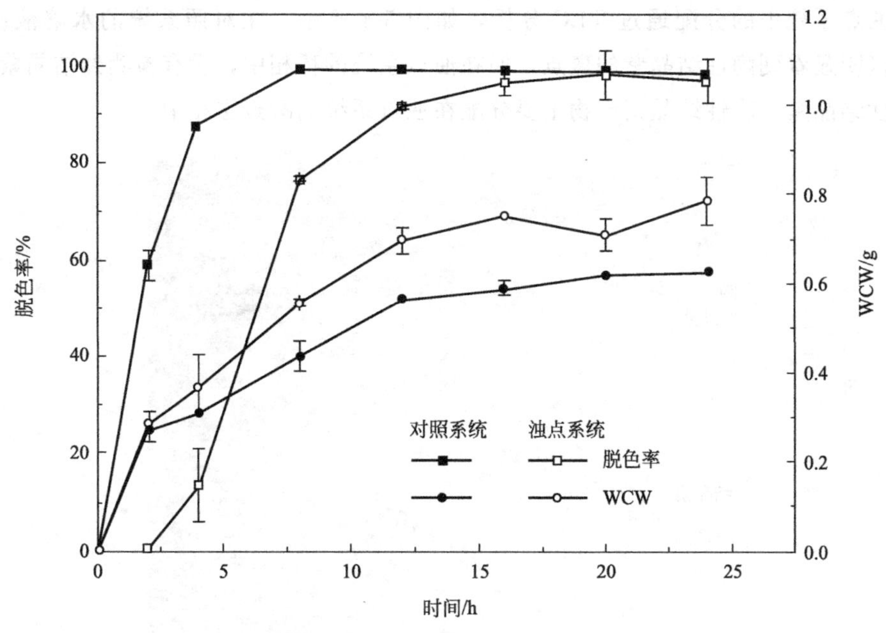

图 5-3　细胞生长及染料脱色曲线

示。浊点系统和对照系统中的细胞生长呈现出相似的趋势。培养 16h 后，细胞生长进入平稳期。浊点系统比对照系统的细胞量更高，说明浊点系统从某种程度上解除了结晶紫对细胞的毒性。浊点系统中，在初始的 16h 内，其脱色率低于对照系统。但随着培养时间的延长，浊点系统中染料脱色率与对照系统基本相同。

浊点系统中，细胞主要分配在稀相中。这使得细胞所处的环境类似于常规的水溶液培养体系。因此，由于脱色过程中水相中有相当一部分结晶紫被增溶到胶束中，水相中的染料浓度相对较低，导致染料的细胞毒性相应地降低。在先前的研究中，我们也发现了类似的结果[69]。同样由于染料增溶到胶束中，结晶紫的生物利用度下降，导致细胞吸附及转化效率降低，因此前 16h，浊点系统的染料脱色率较低。但随着水相中结晶紫被生物利用，其浓度逐渐降低，结晶紫从凝聚层相进一步分配到水相当中并被菌体吸附或转化。最终，浊点系统中的染料脱色率与对照系统基本持平。

5.3.4　染料及其代谢产物在浊点系统中的分配

前文已经分析得知结晶紫的嗜水气单胞菌 DN322p 脱色产物为隐性结晶紫。

在浊点系统中的结晶紫萃取微生物脱色之后，残留的结晶紫及产生的隐性结晶紫在浊点系统中的分配通过 TLC 分析，如图 5-4 所示。在对照系统的水溶液中，可以明显看到隐性结晶紫的斑点，而在浊点系统的稀相中，没有检测到结晶紫和隐性结晶紫。隐性结晶紫产物主要分配在浊点系统的凝聚层相中。

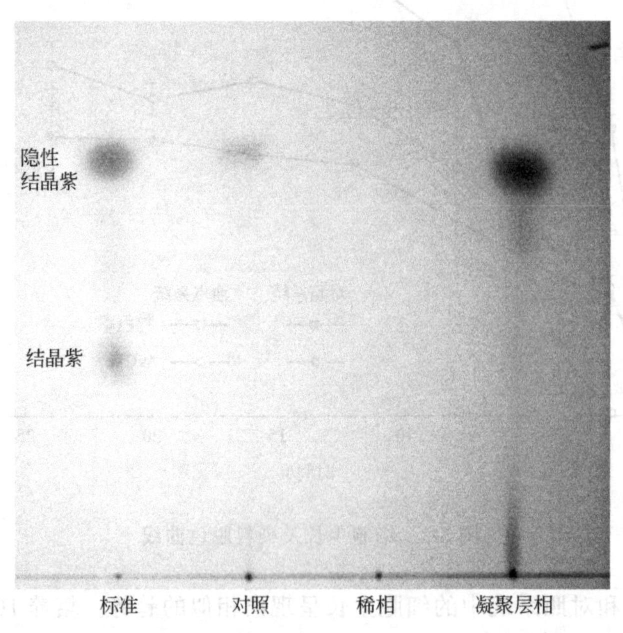

图 5-4　结晶紫及其代谢产物在浊点系统中的分配

TLC 实验说明了萃取微生物脱色后，水相中已无法检测到结晶紫或隐性结晶紫的残留。这将有利于水相的环境排放。

5.3.5　四种三苯基甲烷染料在浊点系统中的萃取微生物脱色

除结晶紫外，实验进一步分析了另外三种三苯基甲烷染料在浊点系统中的萃取微生物脱色情况，如图 5-5 所示。结晶紫、孔雀石绿和灿烂绿在浊点系统中的脱色率与对照系统相当。但是，由于嗜水气单胞菌 DN322p 主要通过吸附方式脱色乙基紫而无法将其转化为无色的代谢产物。因此，乙基紫在浊点系统中的染料脱色率较低。

尽管在传统的水溶液中，有大量关于三苯基甲烷染料微生物脱色的报道[83]。但真正能将此类染料降解为二氧化碳和水的微生物还很稀有[85,88,92]。多数微生物都无法将这些染料彻底脱毒降解[87,89,90]，这极大地限制了这些微生物的工业

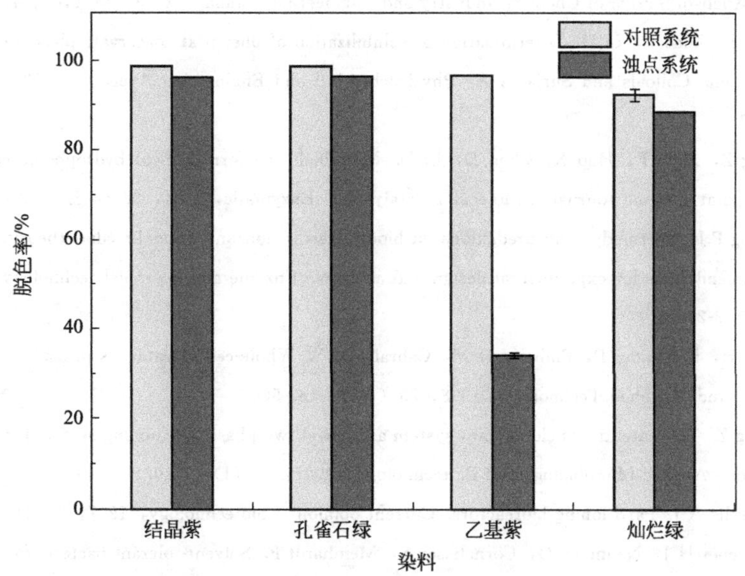

图 5-5 四种三苯基甲烷染料在浊点系统中的萃取微生物脱色

应用。

本章运用介质工程的原理，将浊点系统用于三苯基甲烷染料的微生物脱色中。在浊点系统萃取微生物转化[36]和萃取发酵[69]的基础上，成功实验了染料的萃取微生物脱色。浊点系统中的萃取微生物脱色，合理地优化系统可同时保证染料的脱色率及水相的脱毒。稀相的低毒特点有利于其进一步的污水排放。因此，浊点系统中的染料萃取微生物脱色为三苯基甲烷染料废水的处理提供了一条新的途径。

参 考 文 献

[1] Schick M J. Nonionic surfactants: physical chemistry. Boca Raton: CRC Press, 1987.

[2] Meyer D. Surfaces, interfaces, and colloids: principles and application. New York: John Wiley and Sons, 1999.

[3] Haddou B, Canselier J P, Gourdon C. Cloud point extraction of phenol and benzyl alcohol from aqueous stream. Separation and purification Technology, 2006, 50 (1): 114-121.

[4] Quina F H, Hinze W L. Surfactant-mediated cloud point extractions: An environmentally benign alternative separation approach. Industrial and Engineering Chemistry Research, 1999, 38 (11): 4150-4168.

[5] Hinze W L, Armstrong D W. Organized surfactant assemblies in separation science. ACS Symposium, 1987, 342: 2-82.

[6] Wang Z, Zhao F, Li D. Solubilization of phenol in coacervate phase of cloud point extraction. Hua-

gong Xuebao/Journal of Chemical Industry and Engineering (China), 2003, 54 (10): 1387-1390.

[7] Wang Z, Zhao F, Li D. Determination of solubilization of phenol at coacervate phase of cloud point extraction. Colloids and Surfaces A: Physicochemical and Engineering Aspects, 2003, 216 (1-3): 207-214.

[8] Wang Z, Zhao F, Hao X, Chen D, Li D. Microbial transformation of hydrophobic compound in cloud point system. Journal of Molecular Catalysis B: Enzymatic, 2004, 27 (4-6): 147-153.

[9] Halling P J. Thermodynamic predictions for biocatalysis in nonconventional media: theory, tests, and recommendations for experimental design and analysis. Enzyme and microbial technology, 1994, 16 (3): 178-206.

[10] León R, Fernandes P, Pinheiro H M, Cabral J M S. Whole-cell biocatalysis in organic media. Enzyme And Microbial Technology, 1998, 23 (7-8): 483-500.

[11] Wang Z. The potential of cloud point system as a novel two-phase partitioning system for biotransformation. Applied Microbiology and Biotechnology, 2007, 75 (1): 1-10.

[12] Daugulis A J. Partitioning bioreactors. Current opinion in biotechnology, 1997, 8 (2): 169-174.

[13] Heipieper H J, Neumann G, Cornelissen S, Meinhardt F. Solvent-tolerant bacteria for biotransformations in two-phase fermentation systems. Applied microbiology and biotechnology, 2007, 74 (5): 961-973.

[14] Laane C, Boeren S, Vos K, Veeger C. Rules for optimization of biocatalysis in organic-solvents. Biotechnology and Bioengineering, 1987, 30 (1): 81-87.

[15] Malinowski J J. Two-phase partitioning bioreactors in fermentation technology. Biotechnology advances, 2001, 19 (7): 525-538.

[16] Nikolova P, Ward O P. Whole cell biocatalysis in nonconventional media. Journal of industrial microbiology, 1993, 12 (2): 76-86.

[17] Straathof A J J. Auxiliary phase guidelines for microbial biotransformations of toxic substrate into toxic product. Biotechnology Progress, 2003, 19 (3): 755-762.

[18] Allen C C, Boyd D R, Hempenstall F, Larkin M J, Sharma N D. Contrasting effects of a nonionic surfactant on the biotransformation of polycyclic aromatic hydrocarbons tocis-dihydrodiols by soil bacteria. Applied and Environmental Microbiology, 1999, 65 (3): 1335-1339.

[19] Mata-Sandoval J C, Karns J, Torrents A. Influence of rhamnolipids and Triton X-100 on the biodegradation of three pesticides in aqueous phase and soil slurries. Journal of Agricultural and Food Chemistry, 2001, 49 (7): 3296-3303.

[20] Berti D, Randazzo D, Briganti F, Scozzafava A, Di Gennaro P, Galli E, Bestetti G, Baglioni P. Nonionic micelles promote whole cell bioconversion of aromatic substrates in an aqueous environment. Langmuir, 2002, 18 (16): 6015-6020.

[21] Schmid A, Kollmer A, Mathys R G, Witholt B. Developments toward large-scale bacterial bioprocesses in the presence of bulk amounts of organic solvents. Extremophiles, 1998, 2 (3): 249-256.

[22] Fernandes P, Vidinha P, Ferreira T, Silvestre H, Cabral J M S, Prazeres D M F. Use of free and immobilized Pseudomonas putida cells for the reduction of a thiophene derivative in organic media.

Journal of Molecular Catalysis B: Enzymatic, 2002, 19: 353-361.

[23] Rosche B, Breuer M, Hauer B, Rogers P L. Cells of *Candida utilis* for in vitro (*R*)-phenylacetylcarbinol production in an aqueous/octanol two-phase reactor. Biotechnology letters, 2005, 27 (8): 575-581.

[24] Nikolova P, Ward O P. Whole cell yeast biotransformations in two-phase systems: Effect of solvent on product formation and cell structure. Journal of Industrial Microbiology, 1992, 10 (3-4): 169-177.

[25] Cruz A, Fernandes P, Cabral J M S, Pinheiro H M. Whole-cell bioconversion of β-sitosterol in aqueous-organic two-phase systems. Journal of Molecular Catalysis B: Enzymatic, 2001, 11 (4-6): 579-585.

[26] Cruz A, Fernandes P, Cabral J M S, Pinheiro H M. Solvent partitioning and whole-cell sitosterol bioconversion activity in aqueous-organic two-phase systems. Enzyme And Microbial Technology, 2004, 34 (3-4): 342-353.

[27] Etschmann M M W, Schrader J. An aqueous-organic two-phase bioprocess for efficient production of the natural aroma chemicals 2-phenylethanol and 2-phenylethylacetate with yeast. Applied Microbiology And Biotechnology, 2006, 71 (4): 440-443.

[28] Sinha J, Dey P K, Panda T. Aqueous two-phase: The system of choice for extractive fermentation. Applied Microbiology and Biotechnology, 2000, 54 (4): 476-486.

[29] Zijlstra G M, de Gooijer C D, Tramper J. Extractive bioconversions in aqueous two-phase systems. Current Opinion In Biotechnology, 1998, 9 (2): 171-176.

[30] Jiang Y, Xia H, Guo C, Mahmood I, Liu H. Enzymatic hydrolysis of penicillin in mixed ionic liquids/water two-phase system. Biotechnology Progress, 2007, 23 (4): 829-835.

[31] Wang Z, Zhao F, Hao X, Chen D, Li D. Model of bioconversion of cholesterol in cloud point system. Biochemical Engineering Journal, 2004, 19 (1): 9-13.

[32] Wang Z, Zhao F, Chen D, Li D. Cloud point system as a tool to improve the efficiency of biotransformation. Enzyme and Microbial Technology, 2005, 36 (4): 589-594.

[33] Wang L, Wang Z, Xu J H, Bao D, Qi H. An eco-friendly and sustainable process for enzymatic hydrolysis of penicillin G in cloud point system. Bioprocess and Biosystems Engineering, 2006, 29 (3): 157-162.

[34] Wang Z, Zhao F, Chen D, Li D. Biotransformation of phytosterol to produce androsta-diene-dione by resting cells of *Mycobacterium* in cloud point system. Process Biochemistry, 2006, 41 (3): 557-561.

[35] Wang Z, Wang L, Xu J H, Bao D, Qi H. Enzymatic hydrolysis of penicillin G to 6-aminopenicillanic acid in cloud point system with discrete countercurrent experiment. Enzyme and Microbial Technology, 2007, 41 (1-2): 121-126.

[36] Wang Z, Xu J H, Chen D. Whole cell microbial transformation in cloud point system. Journal of Industrial Microbiology and Biotechnology, 2008, 35 (7): 645-656.

[37] 王志龙. 萃取微生物转化. 北京: 化学工业出版社, 2012.

[38] Wang Z, Xu J H, Wang L, Bao D, Qi H. Thermodynamic equilibrium control of the enzymatic hydrolysis of penicillin G in a cloud point system without pH control. Industrial and Engineering Chemistry Research, 2006, 45 (24): 8049-8055.

[39] Wang Z, Xu J H, Zhang W, Zhuang B, Qi H. In situ extraction of polar product of whole cell microbial transformation with polyethylene glycol-induced cloud point system. Biotechnology Progress, 2008, 24 (5): 1090-1095.

[40] Flygare S, Larsson P O. Steroid transformation in aqueous two-phase systems: side-chain degradation of cholesterol by *Mycobacterium* sp. Enzyme and microbial technology, 1989, 11 (11): 752-759.

[41] Dias A C P, Cabral J M S, Pinheiro H M. Sterol side-chain cleavage with immobilized *Mycobacterium* cells in water-immiscible organic solvents. Enzyme and microbial technology, 1994, 16 (8): 708-714.

[42] Cabral J M S, Aires-Barros M R, Pinheiro H, Prazeres D M F. Biotransformation in organic media by enzymes and whole cells. Journal of biotechnology, 1997, 59 (1-2): 133-143.

[43] Fernandes P, Cruz A, Angelova B, Pinheiro H M, Cabral J M S. Microbial conversion of steroid compounds: recent developments. Enzyme and microbial technology, 2003, 32 (6): 688-705.

[44] Kanda T, Miyata N, Fukui T, Kawamoto T, Tanaka A. Doubly entrapped baker's yeast survives during the long-term stereoselective reduction of ethyl 3-oxobutanoate in an organic solvent. Applied microbiology and biotechnology, 1998, 49 (4): 377-381.

[45] Griffin D R, Gainer J L, Carta G. Asymmetric ketone reduction with immobilized yeast in hexane: biocatalyst deactivation and regeneration. Biotechnology progress, 2001, 17 (2): 304-310.

[46] Wendhausen Jr R, Moran P J, Joekes I, Rodrigues J A R. Continuous process for large-scale preparation of chiral alcohols with baker's yeast immobilized on chrysotile fibers. Journal of Molecular Catalysis B: Enzymatic, 1998, 5 (1-4): 69-73.

[47] Chin-Joe I, Haberland J, Straathof A J, Jongejan J A, Liese A, Heijnen J J. Reduction of ethyl 3-oxobutanoate using non-growing baker's yeast in a continuously operated reactor with cell retention. Enzyme and microbial technology, 2002, 31 (5): 665-672.

[48] Rogers R S, Hackman J R, Mercer V, DeLancey G B. Acetophenone tolerance, chemical adaptation, and residual bioreductive capacity of non-fermenting baker's yeast (*Saccharomyces cerevisiae*) during sequential reactor cycles. Journal of Industrial Microbiology and Biotechnology, 1999, 22 (2): 108-114.

[49] Wang Z, Xu J H, Wang L, Zhang W, Zhuang B, Qi H. Improvement the tolerance of baker's yeast to toxic substrate/product with cloud point system during the whole cell microbial transformation. Enzyme and Microbial Technology, 2007, 41 (3): 296-301.

[50] Rogers P L, Shin H S, Wang B. Biotransformation for L-ephedrine production // Biotreatment, Downstream Processing and Modelling. Berlin: Springer, 1997: 33-59.

[51] Tripathi C M, Agarwal S C, Basu S K. Production of L-phenylacetylcarbinol by fermentation. Journal of fermentation and bioengineering, 1997, 84 (6): 487-492.

[52] Rosche B, Breuer M, Hauer B, Rogers P L. Biphasic aqueous/organic biotransformation of acetalde-

hyde and benzaldehyde by Zymomonas mobilis pyruvate decarboxylase. Biotechnology and bioengineering, 2004, 86 (7): 788-794.

[53] Sandford V, Breuer M, Hauer B, Rogers P, Rosche B. (*R*)-phenylacetylcarbinol production in aqueous/organic two-phase systems using partially purified pyruvate decarboxylase from *Candida utilis*. Biotechnology and bioengineering, 2005, 91 (2): 190-198.

[54] Shin H S, Rogers P L. Biotransformation of benzeldehyde to L-phenylacetylcarbinol, an intermediate in L-ephedrine production, by immobilized *Candida utilis*. Applied microbiology and biotechnology, 1995, 44 (1-2): 7-14.

[55] Long A, Ward O P. Biotransformation of aromatic aldehydes by *Saccharomyces cerevisiae*: investigation of reaction rates. Journal of industrial microbiology, 1989, 4 (1): 49-53.

[56] Zhang W, Wang Z, Li W, Zhuang B, Qi H. Production of L-phenylacetylcarbinol by microbial transformation in polyethylene glycol-induced cloud point system. Applied Microbiology and Biotechnology, 2008, 78 (2): 233-239.

[57] Inoue A, Horikoshi K. A Pseudomonas thrives in high concentrations of toluene. Nature, 1989, 338 (6212): 264-266.

[58] Burton S G, Cowan D A, Woodley J M. The search for the ideal biocatalyst. Nature biotechnology, 2002, 20 (1): 37-45.

[59] Sinensky M. Homeoviscous adaptation—a homeostatic process that regulates the viscosity of membrane lipids in *Escherichia coli*. Proceedings of the National Academy of Sciences, 1974, 71 (2): 522-525.

[60] Heipieper H J, Loffeld B, Keweloh H, de Bont J A. The *cis/trans* isomerisation of unsaturated fatty acids in *Pseudomonas putida* S12: an indicator for environmental stress due to organic compounds. Chemosphere, 1995, 30 (6): 1041-1051.

[61] Kühn I. Alcoholic fermentation in an aqueous two-phase system. Biotechnology and Bioengineering, 1980, 22 (11): 2393-2398.

[62] Cull S G, Holbrey J D, Vargas-Mora V, Seddon K R, Lye G J. Room-temperature ionic liquids as replacements for organic solvents in multiphase bioprocess operations. Biotechnology and bioengineering, 2000, 69 (2): 227-233.

[63] Watanabe H, Tanaka H. A non-ionic surfactant as a new solvent for liquid-liquid extraction of zinc (Ⅱ) with 1-(2-pyridylazo)-2-naphthol. Talanta, 1978, 25 (10): 585-589.

[64] Bordier C. Phase separation of integral membrane proteins in Triton X-114 solution. Journal of Biological Chemistry, 1981, 256 (4): 1604-1607.

[65] Bradoo S, Saxena R K, Gupta R. Two acidothermotolerant lipases from new variants of *Bacillus* spp. World Journal of Microbiology and Biotechnology, 1999, 15 (1): 87-91.

[66] Rathi P, Saxena R K, Gupta R. A novel alkaline lipase from Burkholderia cepacia for detergent formulation. Process Biochemistry, 2001, 37 (2): 187-192.

[67] 王志龙．两相分配生物反应器——浊点系统在生物转化中的应用．中国工程科学，2005，7 (5): 73-78.

[68] Stark D, von Stockar U. In situ product removal (ISPR) in whole cell biotechnology during the last twenty years//Process Integration in Biochemical Engineering. Berlin: Springer, 2003: 149-175.

[69] Pan T, Wang Z, Xu J H, Wu Z, Qi H. Extractive fermentation in cloud point system for lipase production by *Serratia marcescens* ECU1010. Applied Microbiology and Biotechnology, 2010, 85 (6): 1789-1796.

[70] Singh S, Banerjee U C. Enantioselective hydrolysis of methoxyphenyl glycidic acid methyl ester [(±)-MPGM] by a thermostable and alkalostable lipase from *Pseudomonas aeruginosa*. Journal of Molecular Catalysis B: Enzymatic, 2005, 36 (1-6): 30-35.

[71] Zhao L L, Xu J H, Zhao J, Pan J, Wang Z L. Biochemical properties and potential applications of an organic solvent-tolerant lipase isolated from *Serratia marcescens* ECU1010. Process Biochemistry, 2008, 43 (6): 626-633.

[72] Hu Z, Zhang X, Wu Z, Qi H, Wang Z. Perstraction of intracellular pigments by submerged cultivation of *Monascus* in nonionic surfactant micelle aqueous solution. Applied Microbiology and Biotechnology, 2012, 94 (1): 81-89.

[73] Hu M, Zhang X, Wang Z. Releasing intracellular product to prepare whole cell biocatalyst for biosynthesis of *Monascus* pigments in water-edible oil two-phase system. Bioprocess and Biosystems Engineering, 2016, 39 (11): 1785-1791.

[74] Xiong X, Zhang X, Wu Z, Wang Z. Coupled aminophilic reaction and directed metabolic channeling to red *Monascus* pigments by extractive fermentation in nonionic surfactant micelle aqueous solution. Process Biochemistry, 2015, 50 (2): 180-187.

[75] Xiong X, Zhang X, Wu Z, Wang Z. Accumulation of yellow *Monascus* pigments by extractive fermentation in nonionic surfactant micelle aqueous solution. Applied Microbiology and Biotechnology, 2015, 99 (3): 1173-1180.

[76] Pan T, Liu C, Zeng X, Xin Q, Xu M, Deng Y, Dong W. Biotoxicity and bioavailability of hydrophobic organic compounds solubilized in nonionic surfactant micelle phase and cloud point system. Environmental Science and Pollution Research, 2017, 24 (17): 14795-14801.

[77] Wang Z, Dai Z. Extractive microbial fermentation in cloud point system. Enzyme and Microbial Technology, 2010, 46 (6): 407-418.

[78] Wang Z. The potential of cloud point system as a novel two-phase partitioning system for biotransformation. Applied Microbiology and Biotechnology, 2007, 75 (1): 1-10.

[79] Pantsyrnaya T, Delaunay S, Goergen J L, Guseva E, Boudrant J. Solubilization of phenanthrene above cloud point of Brij 30: A new application in biodegradation. Chemosphere, 2013, 92 (2): 192-195.

[80] Pan T, Deng T, Zeng X, Dong W, Yu S. Extractive biodegradation and bioavailability assessment of phenanthrene in the cloud point system by Sphingomonas polyaromaticivorans. Applied Microbiology and Biotechnology, 2016, 100 (1): 431-437.

[81] Pan T, Liu C, Zeng X, Xin Q, Xu M, Deng Y, Dong W. Biotoxicity and bioavailability of hydrophobic organic compounds solubilized in nonionic surfactant micelle phase and cloud point system. En-

vironmental Science and Pollution Research, 2017, 24 (17): 14795-14801.

[82] Gregory P. Dyes and dye intermediates. Kroschwitz J I. Encyclopedia of Chemical Technology: Vol 8. New York: John Wiley and Sons, 1993: 544-545.

[83] Azmi W, Sani R K, Banerjee U C. Biodegradation of triphenylmethane dyes. Enzyme and microbial technology, 1998, 22 (3): 185-191.

[84] Forgacs E, Cserhati T, Oros G. Removal of synthetic dyes from wastewaters: a review. Environment international, 2004, 30 (7): 953-971.

[85] Chen C H, Chang C F, Ho C H, Tsai T L, Liu S M. Biodegradation of crystal violet by a *Shewanella* sp NTOU1. Chemosphere, 2008, 72 (11): 1712-1720.

[86] Yatome C, Yamada S, Ogawa T, Matsui M. Degradation of crystal violet by *Nocardia corallina*. Applied Microbiology and Biotechnology, 1993, 38 (4): 565-569.

[87] Jing W, Jung B G, Kim K S, Lee Y C, Sung N C. Isolation and characterization of *Pseudomonas otitidis* WL-13 and its capacity to decolorize triphenylmethane dyes. Journal of Environmental Sciences, 2009, 21 (7): 960-964.

[88] Chen C C, Liao H J, Cheng C Y, Yen C Y, Chung Y C. Biodegradation of crystal violet by *Pseudomonas putida*. Biotechnology Letters, 2007, 29 (3): 391-396.

[89] Kim M H, Kim Y, Park H J, Lee J S, Kwak S N, Jung W H, Lee S G, Kim D, Lee Y C, Oh T K. Structural insight into bioremediation of triphenylmethane dyes by *Citrobacter* sp. triphenylmethane reductase. Journal of Biological Chemistry, 2008, 283 (46): 31981-31990.

[90] Jang M S, Lee Y M, Kim C H, Lee J H, Kang D W, Kim S J, Lee Y C. Triphenylmethane reductase from *Citrobacter* sp. strain KCTC 18061P: purification, characterization, gene cloning, and overexpression of a functional protein in Escherichia coli. Applied and environmental microbiology, 2005, 71 (12): 7955-7960.

[91] Xue Y, Qian C, Wang Z, Xu J H, Yang R, Qi H. Investigation of extractive microbial transformation in nonionic surfactant micelle aqueous solution using response surface methodology. Applied Microbiology and Biotechnology, 2010, 85 (3): 517-524.

[92] Chen C H, Chang C F, Liu S M. Partial degradation mechanisms of malachite green and methyl violet B by *Shewanella decolorationis* NTOU1 under anaerobic conditions. Journal of Hazardous Materials, 2010, 177 (1-3): 281-289.

第6章

非离子表面活性剂的回收

6.1
表面活性剂的回收方法

6.1.1 常规方法

无论是浊点萃取还是浊点系统中的萃取微生物转化，分离有机物并回收表面活性剂都不可或缺[1]。这一方面关系到生产成本，另一方面也与工业过程的环境影响相关。然而，非离子表面活性剂的下游加工过程的选择是由其上游加工过程决定的。浊点系统的主要组成介质是非离子表面活性剂，一方面表面活性剂对萃取对象或者微生物转化产物具有浓缩和部分纯化作用，有利于下游加工。另一方面，表面活性剂的引入也增加了反应体系的新组分，使得分离过程变得更加复杂。下游加工的关键是目标物与非离子表面活性剂的分离[2]。

但目前，有关产物分离和表面活性剂回收的方法还较少。非离子表面活性剂具有高沸点，因此当增溶物为低沸点有机物时，可采用蒸发的方法进行分离[3]。而对于可电离的有机物，调节体系 pH 值也能实现化合物和表面活性剂的分离[4]。回收蛋白质[5,6]、金属离子[7]及通过添加酸性或碱性基团的方法回收有机物[8]的研究已见诸报道。然而，如何从非离子表面活性剂中回收非挥发性有机物仍然具有挑战性[9]。受污染表面活性剂分离方法的优缺点如表 6-1 所示。

表 6-1 受污染表面活性剂分离方法的优缺点[9]

方法	适用范围	优点	缺点
气提	挥发性有机物	高效,设备要求低	大量泡沫,不适合高浓度表面活性剂
真空抽提	半挥发有机物	适合高浓度表面活性剂,适合低挥发有机物	大量泡沫,高消耗,需要真空环境,需要加热
渗透蒸发	挥发性有机物	避免泡沫	设备要求高,昂贵
溶剂萃取	有机物	对不挥发有机物有效,高效	溶剂花费多,溶剂需要处理

续表

方法	适用范围	优点	缺点
加入抗衡离子形成表面活性剂沉淀	离子表面活性剂	低价,操作方便	抗衡离子需要进一步处理
低温形成表面活性剂沉淀	Krafft 点接近室温的离子表面活性剂	低价,操作方便,回收的表面活性剂可再利用	表面活性剂的 Krafft 点必须接近室温
用离子交换树脂吸附	离子表面活性剂	同时移除表面活性剂和有机物,对高浓度表面活性剂有效	树脂花费多,无法回收表面活性剂
活性炭吸附	预处理前的抛光步骤	同时移除表面活性剂和有机物,对所有表面活性剂和有机物有效	活性炭花费多,不适用于高浓度表面活性剂
反胶束萃取	高疏水性有机物	同时移除表面活性剂和有机物	难以将表面活性剂与有机物分离
基于交换阳离子的相过渡	超低界面张力和多价阳离子的阴离子表面活性剂	有效将 Winsor Ⅲ微乳液转化为 Winsor Ⅰ微乳液	仅限于 Winsor Ⅲ微乳液体系,无法将表面活性剂与有机物分离

6.1.2 Winsor 微乳液回收非离子表面活性剂

非挥发性非离子表面活性剂是一种环境友好的溶剂[10]，使得浊点系统微生物转化过程中的产物回收不能像传统的水-有机溶剂两相萃取那样，照搬常规的蒸发过程。除了烦琐的色谱分离外，只有很少的报道涉及从特殊目标化合物中大规模分离非离子表面活性剂[11]。例如，与只能提取中性溶质的水相两相系统一样[12]，已经提出了调节浊点系统的 pH 值以利用酸性或碱性部分反萃溶质的方法[13]。也有文献记录了通过全蒸发从表面活性剂溶液中回收挥发性有机化合物的情况。在蛋白质的浊点萃取后，将共聚物 EOPO 添加到凝聚层中形成了一个新的两相系统，其中蛋白质由于其强排阻作用而分配到水相中[14]。这种现象已被用于从非离子表面活性剂中分离目标蛋白[15]。Winsor Ⅱ微乳液已成功用于从非离子表面活性剂溶液中分离亲水性胆固醇氧化酶[5]。发展产物分离和非离子表面活性剂回收的通用策略对于浊点系统的工业应用是至关重要的。

1986 年，报道了一种三相系统，它包括水、有机溶剂和表面活性剂[16]。有时，需要添加有机醇等小分子作为助溶剂。这种三相系统后来被称作 Winsor 微乳液。如图 6-1 所示，Winsor 微乳液根据相组成的不同，分为 Winsor Ⅰ、Winsor Ⅱ和 WinsorⅢ型三种形式。其中 Winsor Ⅰ微乳液由水包油（O/W）相

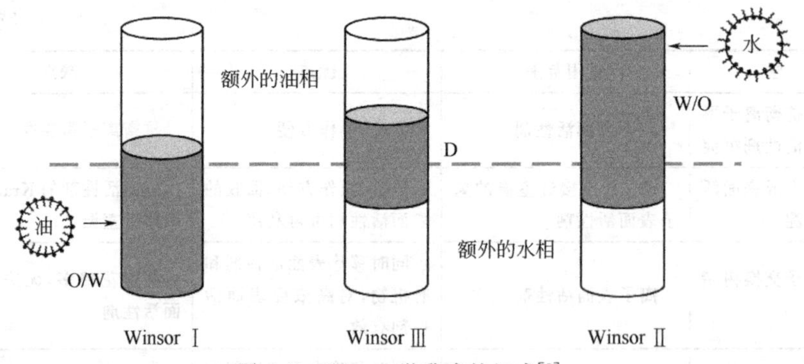

图 6-1 Winsor 微乳液的组成[2]

和额外的油相组成，Winsor Ⅱ由油包水（W/O）相和额外的水相组成，Winsor Ⅲ微乳液由额外的油相、额外的水相和双连续（D）相组成。有时，水、表面活性剂和有机溶剂形成单相，也被称为 Winsor Ⅳ。油和水的组成、温度和体积比是决定微乳液类型的重要因素，见表 6-2[17]。有机溶剂极性的影响如表 6-2第 1、3、5 号所示，有机溶剂由异丁醇、乙酸丁酯变为乙醚，微乳液由 Winsor Ⅱ转为 Winsor Ⅰ；非离子表面活性剂的 HLB 值的影响如表 6-2 第 10、7 和 9 号所示，非离子表面活性剂由 Triton X-45 改为 Triton X-114 和 Triton X-100，微乳液由 Winsor Ⅲ改为 Winsor Ⅰ；温度的影响如表 6-2 第 2、4 和 6 号所示，表面活性剂与任何一种有机溶剂都能在 25℃下形成 Winsor Ⅱ微乳液。这些都是确定微乳

表 6-2 复合组分及参数对微乳液类型的影响[17]

序号	表面活性剂	有机溶剂	温度/℃	表面活性剂在水中的体积分数/%	油水比例/(mL∶mL)	表面活性剂在水中的体积分数/%	Winsor
1	Triton X-114	异丁醇	6	10	20∶20	0.5	Ⅱ
2	Triton X-114	异丁醇	25	10	20∶20	0.7	Ⅱ
3	Triton X-114	乙酸丁酯	6	10	20∶20	43.5	Ⅱ
4	Triton X-114	乙酸丁酯	25	10	20∶20	10.7	Ⅱ
5	Triton X-114	乙醚	6	10	20∶20	81.7	Ⅰ
6	Triton X-114	乙醚	25	10	20∶20	30.2	Ⅱ
7	Triton X-114	乙醚	6	10	40∶20	58.5	Ⅰ
8	Triton X-114	乙醚	6	2.5	20∶20	62.8	Ⅰ
9	Triton X-100	乙醚	6	10	20∶20	91.7	Ⅰ
10	Triton X-45	乙醚	6	10	20∶20	2.05	Ⅲ

液类型的重要因素。通过选择亲水性非离子表面活性剂、相对极性的有机溶剂，在相对低温的条件下，可以形成 Winsor Ⅰ微乳液。如果将微生物转化产物溶解在有机溶剂中，则可在浊点系统中在微生物转化的下游过程中将产物和非离子表面活性剂分离。

作为一种重要的分离纯化介质，甚至有人提议将微乳液作为一种液相萃取的吸附剂来利用[18]。如图 6-2 所示，Liang 等利用 Winsor Ⅰ微乳液成功将高沸点、中等极性的有机物从表面活性剂中分离出来，从而实现了有机物和非离子表面活性剂的回收[1]。而相对亲水的大分子蛋白脂肪酶可以用 Winsor Ⅱ微乳液进行分离与回收[19]。

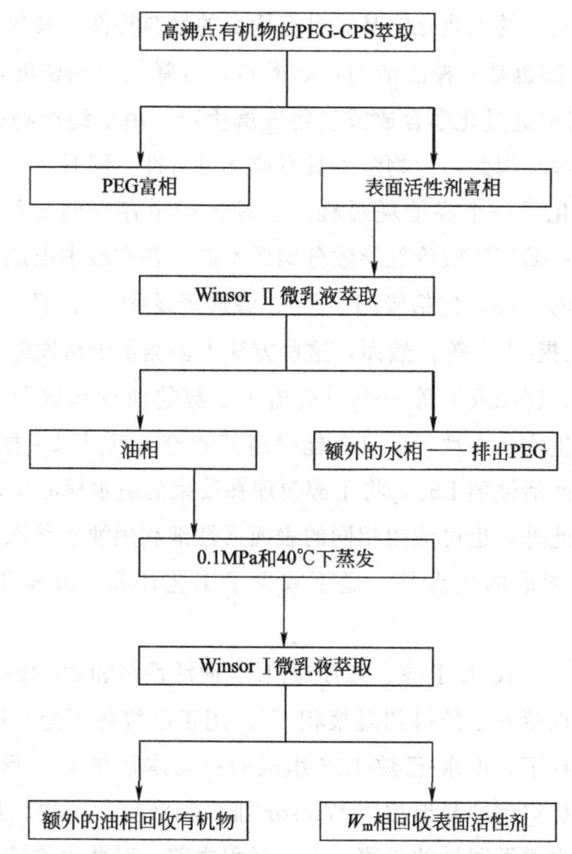

图 6-2 Winsor Ⅰ微乳液回收表面活性剂与高沸点有机物[1]

(1) L-PAC 萃取微生物转化的表面活性剂回收策略

L-PAC 是一种药物中间体，常用于麻黄素和伪麻黄素的合成，也可用于低血压和哮喘的治疗[20]。酿酒酵母将苯甲醛转化为 L-PAC 的微生物转化受到有毒底物和产物的严重抑制[21]。此外，L-PAC 具有高极性，lg P 达到了 0.6。一般情况下，高极性代谢产物很难在水-有机溶剂两相系统中进行萃取微生物转化，因为极性产物的生物相容性和原位去除无法同时实现。

以全细胞酿酒酵母[22]在浊点体系中将苯甲醛微生物转化为 L-PAC 为例，建立了微乳液萃取分离产品和回收表面活性剂的通用策略[17]。首先使用 Winsor Ⅰ微乳液成功分离微生物转化液中的产物和非离子表面活性剂，然后用 Winsor Ⅱ微乳液回收非离子表面活性剂。在单级 Winsor Ⅰ微乳液萃取过程中，产品回收率为 76.9%，非离子表面活性剂回收率为 66.5%[17]。

(2) 丁醇萃取发酵的表面活性剂回收策略

丁醇是一种重要的工业化学品，具有优良的燃料特性。其热值比乙醇高，冰点较低[23-25]。丁醇也是一种清洁的生物燃料，可降低烟气密度，有助于保护环境[24,25]。丁醇可以通过化学合成或生物发酵生产。由于终产物的高毒性严重阻碍了发酵生产丁醇，因此，生物法不具有商业可行性。同时，发酵液中丁醇浓度较低，其分离纯化是一个高能耗过程。丁醇生产中存在两个相互关联的挑战：①最终产品毒性；②丁醇效价低导致分离成本高。在文献中报道了各种方法来减弱丁醇对微生物的毒性，包括吸附[26-28]、渗透蒸发[29-31]、超临界萃取[32]、液-液萃取[33,34]和汽提[35,36]等。然而，这些方法大多侧重于从发酵完成的培养液中分离丁醇。最近，Dhamole 等[3,37,38]提出了丁醇的浊点萃取发酵系统，它不仅能减弱丁醇对微生物的毒性，而且还能提高其效价（反过来又能提高产量）。作者采用非离子表面活性剂 L62，将丁醇包埋在胶束的疏水核心中，从而从发酵液中分离出丁醇。此外，也可使用相同的表面活性剂利用浊点萃取从发酵液中分离丁醇。使用浊点萃取的优点是，显著减少了工艺体积，并将丁醇体积降低到 1/6～1/4。

Raut 等在由 54.4g/L 丁醇、127.1g/L 表面活性剂和 80.4g/L 水组成的模型体系中，通过浊点萃取丁醇得到凝聚相[39]。用正己烷将丁醇从凝聚相反萃至有机相。在丁醇存在下，由水-己烷-L62 组成的三元体系在 45℃形成 Winsor Ⅲ微乳液。浊点萃取后得到的凝聚相用 Winsor Ⅲ微乳液反萃丁醇。从微生物转化发酵液中分离丁醇也遵循同样的步骤。在此过程之前，对微生物转化培养基进行真空过滤去除细胞，得到澄清的发酵液。在丁醇发酵过程中添加的 L62 用于从微生物转化液中分离丁醇[3]。发酵液中含有 10g/L 丁醇和 3%的 L62。为了模拟丁醇产生菌的生长条件，并预测 L62 能使丁醇产量增加 1 倍，将丁醇浓度控制在 40g/L，并将丁醇浓缩到凝聚相。在 50℃孵育 30min 后得到含有凝聚层相和稀相的两相系统。

丁醇在凝聚层相和水相中的分布分别为 31.8%和 68.2%（质量分数）[39]。丁醇的分布取决于表面活性剂对丁醇的捕收能力。L62 对丁醇的捕收能力为 0.4g/g 表面活性剂。由此可以看出，分配到凝聚层相的丁醇与 L62 的丁醇捕收能力相匹配。发酵液体积减小（初始发酵液体积/凝聚相体积=4.35），下游分离工艺体积减小，从而显著降低了丁醇提取的能耗。

使用 Winsor Ⅲ微乳液对丁醇和 L62 进行反萃以分离凝聚层[39]。将从发酵液中得到的凝聚层与己烷以 1∶1 的体积比混合，并在 45℃下孵育。在三个阶段中，丁醇的分布在额外的油相中为 49.1%，在双连续相中为 0%，在额外的水相中为 50.62%，而 L62 在额外的油相中为 0%，在双连续相中为 58.9%，在额外的水相中为 38.9%。最后，对 Winsor Ⅲ微乳液的各相进行丁醇、L62 和己烷的回收。额外的油相中没有表面活性剂，因此，蒸发己烷后，分离出丁醇。可以将己烷回收再利用。双连续相仅包含表面活性剂和水，因此可以直接循环至发酵液中。额外的水相含有丁醇、水和表面活性剂，通过蒸发该相中的丁醇和水，回收了表面活性剂。这样回收的表面活性剂被再次用于从模型系统中提取丁醇，表面活性剂的再利用进行了三个周期。首次运行后，L62 的总回收率为 97.9%。在第三次运行期间，体积比分别从第一次和第二次运行期间的 3.9∶1 和 4.0∶1 降低到 2.7∶1。L62 在第一次、第二次和第三次实验中的丁醇浓缩系数分别为 0.429、0.438 和 0.437。从结果中可以看出，丁醇捕收能力即使在三个循环后仍然不受影响。因此，可以得出结论，表面活性剂可以重复用于丁醇的提取[39]。

(3) 脂肪酶萃取发酵的表面活性剂回收策略

脂肪酶（E.C.3.1.1.3）是能够催化三酰甘油水解为甘油和脂肪酸的水解酶[40]。此外，脂肪酶还可催化酯水解、酯交换以及酯合成，并具有对映选择性。脂肪酶的底物（长链三酰甘油）不溶于水，因此，它们首先溶解在有机溶剂中，然后与缓冲液混合形成两相体系用于生物催化。脂肪酶溶于水，但具有一定的疏水性，能在水和有机介质界面催化反应。但有机溶剂可能会使脂肪酶变性并导致其构象变化，从而影响其功能和催化活性。脂肪酶已成为领先的生物催化剂/生物促进剂之一，在食品、洗涤剂、化学和医药工业中越来越受欢迎。

脂肪酶具有多种商业用途，包括生产生物聚合物、生物柴油、药品、农用化学品、化妆品和香料等[40]。脂肪酶由于其稳定性、高选择性和广泛的底物特异性而受到工业界的特别关注。脂肪酶具有催化活性高、反应条件温和、环境友好、化学选择性高、对映选择性好和区域选择性好等优点，是一类具有广泛应用前景的酶。脂肪酶只能在温和的条件下工作，与化学催化剂相比，脂肪酶的活性弱得多，生产成本高且耗时长。

Serratia marcescens ECU1010 的脂肪酶是合成心血管病药物地尔硫䓬前体——(±)-MPGM[*trans*-3-(4′-甲氧基苯基)缩水甘油酸甲酯] 的重要催化剂[41]。笔者团队探索了浊点系统中 *S. marcescens* ECU1010 脂肪酶的萃取发酵。这是浊点系统用于大分子萃取发酵的首例报道[42]。浊点系统由混合非离子表面活性剂的水溶液 Triton X-114 和 Triton X-45（4∶1）组成。萃取发酵使脂肪酶浓度从1700U/L 提高到 2800U/L。在浊点系统中，脂肪酶倾向于分配在凝聚层相，而菌体和其他亲水性的蛋白则分配在稀相。脂肪酶的浓缩因子为 4.2，纯化因子为1.3，这为下游分离过程带来了便利。在浊点系统的凝聚层相中添加不同的醇(异丁醇、2-苯乙醇和 1-辛醇)，研究非离子表面活性剂与和脂肪酶的分离。结果显示，形成 Winsor Ⅱ微乳液的醇-水-非离子表面活性剂三元系统可以将非离子表面活性剂萃取到油相，另一相为富含脂肪酶的水溶液相。特别是以 2-苯乙醇为有机溶剂的 Winsor Ⅱ微乳液萃取，油水界面处没有蛋白-表面活性剂复合物，脂肪酶的回收率达到 80%以上，而且有效地实现了非离子表面活性剂的分离。

Ooi 等采用浊点萃取，从 *Burkholderia* sp. ST8 原始发酵液中初步回收了脂肪酶[43]。作者对不同的非离子表面活性剂溶液进行了筛选，发现在相对较低的浊点温度（9.0～37.5℃）下，脂肪酶对胶束相具有不同的亲和力。在分别由 Pluronic L81 和 TX-114 生成的水-胶束两相系统中，由于表面活性剂胶束更加疏水，脂肪酶被分配到胶束相。由 24%（质量分数）Pluronic L81 组成的浊点系统为脂肪酶分配到胶束底相提供了最佳条件。在水-胶束两相系统中加入 0.5%(质量分数）KCl 进一步改善了脂肪酶的分配，使其分配系数高达 7.2。在萃取了脂肪酶的胶束相中加入新的硫氰酸钾溶液，实现了脂肪酶向新水溶液的转移。此方法可将大多数脂肪酶提取到下部水相，约 80%的表面活性剂分配到小体积的黏性上相。此方法实现了胶束相中脂肪酶的回收，还可进一步改进以进行大规模回收。并且，合适的回收过程让浊点萃取更具有成本效益，因为表面活性剂 Pluronic L81 可循环用于后续脂肪酶的萃取[43]。

6.1.3 表面活性剂与室温离子液体的相互作用

表面活性剂和室温离子液体作为两种绿色溶剂，是取代传统有机溶剂的一种选择，在萃取、分离、纯化和生物转化等很多领域都有深入的探索。近年来，两种溶剂的交叉科学的研究方兴未艾。2005 年，Gao 等研究了 TX-100＋水＋

$[Bmim]PF_6$ 三元系统[44]。如图 6-3 所示，Gao 等发现，此三元微乳液系统有着明显的两相区和单相区。

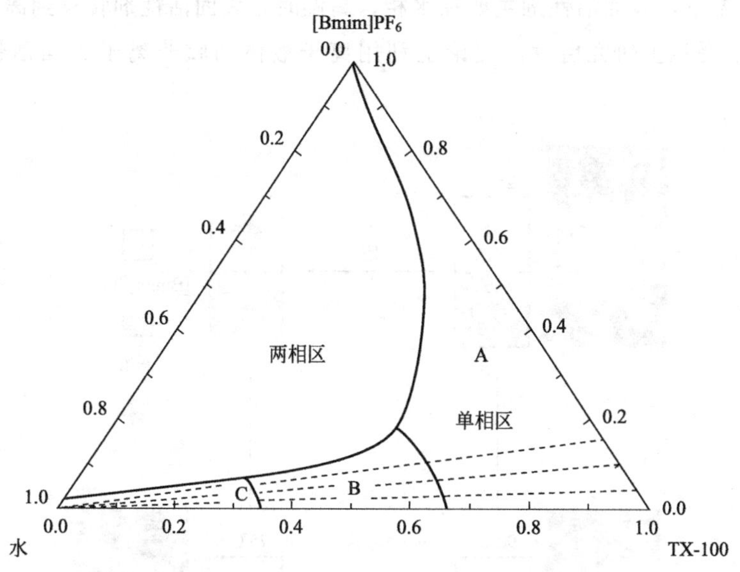

图 6-3　TX-100＋水＋$[Bmim]PF_6$ 三元系统的相图[44]

离子液体是熔点接近室温的盐，由弱 Lewis 酸性阳离子和弱 Lewis 碱性阴离子组成，具有很强的极性。例如 $[Bmim]PF_6$（水不溶性离子液体）比乙腈的极性大，但比甲醇的极性小[45]。有些离子液体可以替代水与表面活性剂形成自组装胶束。在此之前，表面活性剂只能与水或者一些极性有机溶剂形成自组装胶束。因此，离子液体的出现，极大地扩展了表面活性剂胶束自组装的溶剂数量[46]。与非离子表面活性剂胶束水溶液的浊点相似，离子液体中的非离子表面活性剂也表现出相对较低的浊点[47-49]。

水＋[Bmim]Cl＋Triton X-114 三元系统中的两两组分，如水＋[Bmim]Cl、水＋Triton X-114 或者 [Bmim]Cl＋Triton X-114，在室温下可以完全混溶。然而，在三元区存在一个不混溶窗口，且不混溶窗口的面积随温度的升高而增大(称为岛型三元系统)[2]。Álvarez 等公布了 $C_2MIMC_2SO_4$＋Triton X＋水三元系统的相图，发现亲水性比 $C_2MIMC_2SO_4$ 更强的离子液体氯化胆碱和疏水性表面活性剂之间能形成了更大的两相区[50]。两相系统由富离子液相和富非离子表面活性剂相（称为离子液体-非离子表面活性剂水溶液两相系统）组成[51]。这种两相系统包含一种含有带电荷离子液体的表面活性剂稀相，令溶质在三元系统中有了新的分配行为。在氯化胆碱-Tween 80-水三元系统中，双氯芬酸和异丁苯丙酸

(少量水溶性）都能较好地迁移到非离子表面活性剂富相中[52]。

与此同时，He 等发现了胶束在水相和离子液体相间的穿梭现象[53]。主要表现为室温下，表面活性剂主要在水相；高温时，表面活性剂转移到离子液体相（图 6-4)。受以上研究启发，理论上利用离子液体回收非离子表面活性剂是可行的。

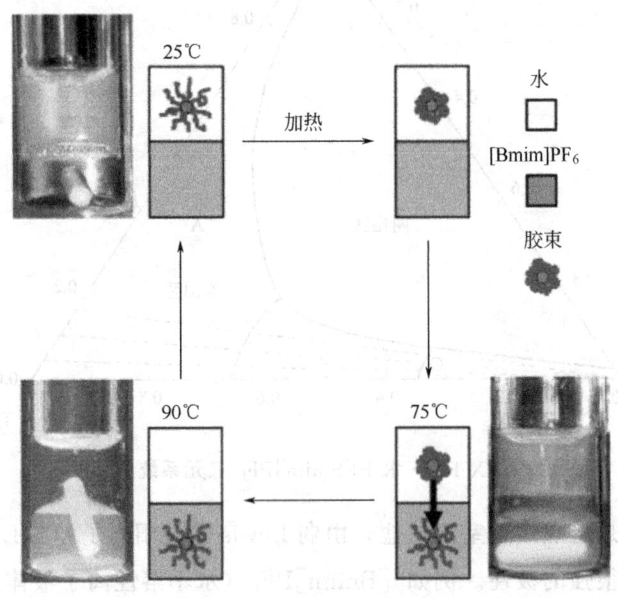

图 6-4　胶束在水相和离子液体相的穿梭[53]

Zhao 等研究了阴离子染料在离子液体-非离子表面活性剂水溶液两相系统中的分配作用[54]。系统中的含水率影响非离子表面活性剂的浊点。低浊点一般对应相对较低的临界非离子表面活性剂浓度。在 70% [Bmim]Cl 水溶液中，最低浊点和临界非离子表面活性剂浓度均显著增大。在 [Bmim]Cl 水溶液中，浊点随离子液体含量的增加，先升高后降低。离子液体的含量为 40%和 70%时，浊点分别达到最高点和最低点。进一步提高离子液体的浓度，则表面活性剂的浊点会继续升高。换言之，在相对较高的离子液体含量下有最低的浊点。Triton X-114在 [Bmpy]Cl 水溶液中的浊点进一步证实了这一结论。此外，还研究了表面活性剂浓度对高浓度 [Bmim]Cl 水溶液中非离子表面活性剂最低浊点的影响。当离子液体含量较高时，浊点随 [Bmim]Cl 含量的增加先减小后增大。最低浊点随非离子表面活性剂浓度从 5%增加到 20%而降低，当表面活性剂浓度为 50%时则升高。值得一提的是，在高非离子表面活性剂浓度为 50%条件下，离子液体含量相对较低（如 40%离子液体含量时的最低浊点），也对应于较高的离子液体与水的比例（4∶1）。

染料麝香草酚蓝在不同 pH 值的 70%乙醇溶液中，离子性质不同[54]。pH 值为 12 时，麝香草酚蓝为阴离子态，溶于水并呈现蓝色，吸收峰在 660nm 处。pH 值为 1.5 时，为两性离子态，呈红色，不溶于水，但溶于乙醇，吸收峰在 550nm 处。在阴离子态和两性离子态之间，麝香草酚蓝在中性（pH 值为 7）时，呈黄色，同时也是一种阴离子态，不溶于水，但溶于乙醇，吸收峰在 370nm 处。5%的 Triton X-114 水溶液在 30℃下形成浊点系统。在 pH 值为 1.5～12 范围内，麝香草酚蓝都分配在凝聚层相。但是，用离子液体-表面活性剂水溶液两相系统的离子液体替代稀相后，染料的分配发生显著变化。两性离子态（pH＝1.5）的麝香草酚蓝为水不溶性，优先分配到表面活性剂富集相。相反，碱性条件下的阴离子麝香草酚蓝（pH＝12）是水溶性的，更倾向于分配在离子液相（底相）。中性条件下的麝香草酚蓝也是一种阴离子，但不溶于水（pH＝7），因此倾向于分配到富含表面活性剂的相。

用 Triton X-114 浓度固定在 25%的溶液进一步测试了相组成对麝香草酚蓝分配的影响[54]。在相同 pH 条件下，分配系数随离子液体含量的增加而增大。与水溶性阴离子麝香草酚蓝的提取一致，在相同的相组成条件下，从酸性到碱性变化过程中，染料的分配系数也会增大。进一步研究了各种水溶性离子染料在[Bmim]Cl＋Triton X-114 双水相系统中的分配。在实验条件下，阳离子染料结晶紫和碱性红呈正电性或电中性状态，尽管在酸性条件下具有较高的水溶性，但在富表面活性剂相和富离子液体相之间几乎均匀分布，排除了高水溶性是离子液体-非离子表面活性剂双水相系统萃取的关键特征。相反，阴离子染料茜素红、氯冉酸和伊文思蓝处于阴离子或中性状态，在碱性条件下表现出较高的分配系数。

在阴离子状态下，富离子液体相优先萃取有机染料[54]。将阴离子染料萃取到富含离子液体的相中可以用可与水混溶的离子液体作为水溶助剂来解释[55]。分子动力学模拟表明，离子液体的大多数阳离子在性质上是有机的，因此可以与有机（疏水）溶质相互作用。因此，离子液体的阳离子和溶质的阴离子之间的静电力非常重要，这与水溶性阳离子染料在离子液体富集相中的有限分配是一致的。此外，离子液体的阳离子和溶质的阴离子之间的相互作用还应与阳离子的非极性部分和溶质分子之间的分散力以及其他特定的相互作用（如氢键相互作用、π-π 相互作用）有关。还观察到，将阴离子物种提取到富离子液体相与离子液体含量、溶质溶解度以及其他溶质特性有关。双氯芬酸和异丁苯丙酸（水溶性有限）都优先分配到氯化胆碱-Tween 80 双水相系统中富含非离子表面活性剂的相中[52]。

新型离子液体-非离子表面活性剂双水相系统提供了一个带电的离子液体水相，与非离子表面活性剂富相共存[54]。高离子液体含量的富离子液体相具有特殊的溶剂性质，可用于阴离子染料的萃取。新型离子液体-非离子表面活性剂双水相系统中阴离子溶质的高分配系数可能有助于将这些阴离子溶质从非离子表面活性剂水溶液中萃取出来。而如何进一步回收离子液体，将是另一个挑战[54]。

6.2 表面活性剂-水-离子液体三元微乳液

前面三章先后讨论了嗜水气单胞菌 DN322p 的染料脱色机理、三苯基甲烷染料在浊点系统中的分配机理及浊点系统中染料的萃取微生物脱色。在此基础上，本章重点讨论如何分离浊点系统凝聚层相中增溶的有机物并回收非离子表面活性剂。

浊点系统作为一种绿色体系可代替传统的两相系统应用于很多领域[56]。如两相催化、萃取微生物转化和萃取发酵、萃取及分离过程[42,57-60]。然而，有关如何从浊点系统中分离出有机物并回收表面活性剂的报道十分有限。

非离子表面活性剂具有高沸点，因此，对于低沸点的有机物，可采用真空蒸发浓缩的方法来实现产物的分离和表面活性剂的回收[61]。浊点系统的凝聚层相对电中性的有机物萃取能力较强。所以，对于可电离的化合物，通过调节 pH 的方法可达到分离产物与表面活性剂的目的[4]。浊点系统中非离子表面活性剂回收的主要难题是如何分离那些高沸点非电离的有机化合物。

微乳液由表面活性剂-水-有机溶剂组成，是一种热力学稳定系统。微乳液通常可以分为 Wisnor Ⅰ、Winsor Ⅱ和 Wisnor Ⅲ三种不同的类型。Winsor Ⅰ微乳液是非离子表面活性剂的水溶液和过剩的有机溶剂组成的两相系统。如果有机溶剂在水-有机溶剂两相系统中具有较高的分配系数，应用 Winsor Ⅰ微乳液理论上可以回收非离子表面活性剂[1]。但此法只对高 HLB 值非离子表面活性剂有效，不利于低 HLB 值非离子表面活性剂的回收。

对于生物大分子，高度亲水性的脂肪酶等可采用 Winsor Ⅱ微乳液萃取。采用 Wisnor Ⅱ微乳液分离胆固醇氧化酶[5]、脂肪酶[61,62]等已有文献报道。在低

HLB 值的 Triton X-114＋Triton X-45 非离子表面活性剂浊点系统中进行脂肪酶的萃取发酵，其下游加工过程选择异丁醇或二苯乙醇作为有机溶剂的 Winsor Ⅱ微乳液萃取，成功将表面活性剂萃取到有机相而脂肪酶保留在水相[19]。

Winsor 微乳液萃取需要添加有机溶剂，这使得浊点系统这种“绿色系统”不再“绿色”。当前，有研究发现在离子液体-水-表面活性剂三相系统中，离子液体表现出有机溶剂的特征[44]。同时，有研究者报道了胶束在离子液体-水两相系统中的穿梭现象[53,63]。并且，离子液体萃取金属离子、苯系物、染料和其他有机无机化合物的研究已十分广泛[64-67]。王志龙发现了有机染料浊点萃取及其离子液体的反萃过程[2]。所有这些结果给我们以启示：用离子液体从浊点系统的凝聚层相中分离有机物并回收非离子表面活性剂是可行的。

6.2.1 确定表面活性剂-水-离子液体三元微乳液两相区边界

表面活性剂-水-离子液体三元微乳液两相区边界采用直接观察法确定。在纯的离子液体中，加少量水，滴加非离子表面活性剂直到出现浑浊，记下溶液组成；再向溶液中加少量水，直至溶液澄清；再次向溶液中滴加非离子表面活性剂直至出现浑浊，记下溶液组成。如此反复，直至体积太大无法继续下去。在三元相图上连接浑浊时各点的溶液组成，得到三元微乳液的两相区边界。

6.2.2 电导法测定单相区微乳液类型

沿三元相图的单相区选一条直线作为研究路线，大致区分离子液体包水（W/IL)、水包离子液体（IL/W）和双连续（bicontinuous）相区。

操作方法：选择离子液体和非离子表面活性剂合适比例的系统，每次加少量水，混匀后用电导仪测定溶液电导率。最后以水含量为横坐标，电导率为纵坐标作图。在电导率的拐点处即是三元微乳液相变点[68]。

6.2.3 染料与非离子表面活性剂的检测

非离子表面活性剂 Triton X-114 通过 HPLC 检测。检测条件为：检测波长为 230nm；流动相为 95％乙腈＋磷酸缓冲液；流速为 1mL/min；出峰时间为 2.33min。

实验选用五种染料，分别为蒽醌染料——茜素，偶氮染料——苋菜红和甲基橙，三苯基甲烷染料——结晶紫和孔雀石绿。采用分光光度法检测，其基本性质如表 6-3 所示。lg *P* 值可在网站 http：//www.chemicalize.org 查询得到。

表 6-3　实验用染料的基本性质

类型	染料	分子结构	lg *P*	吸收峰/nm
蒽醌	茜素	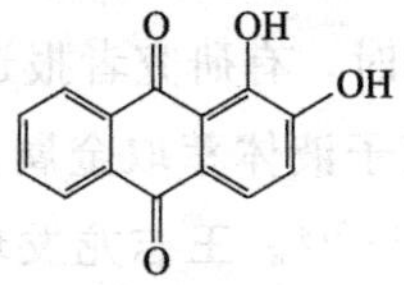	2.96	462(pH＜7) 570(pH＞7)
偶氮	苋莱红	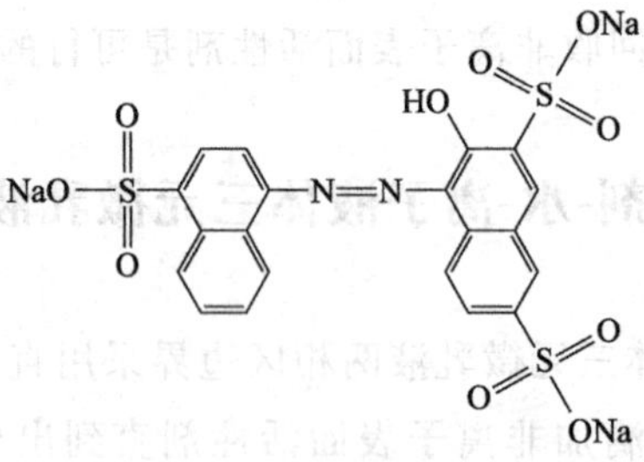	−1.98	520
	甲基橙	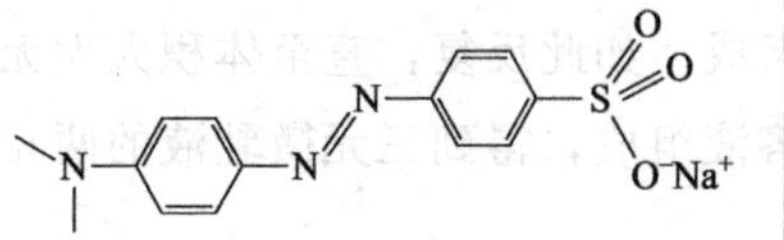	3.47	463
三苯基甲烷	结晶紫	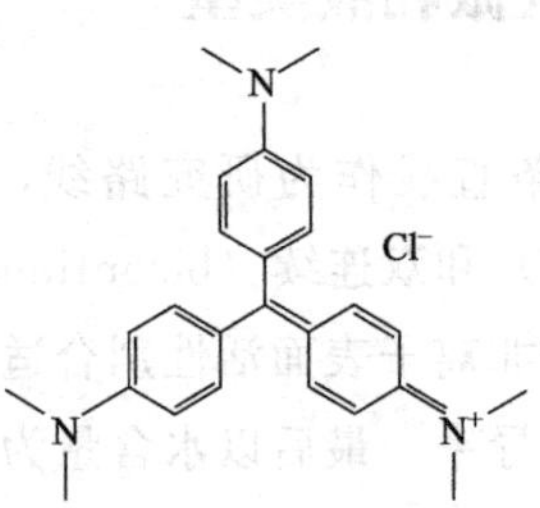	1.77	590
	孔雀石绿	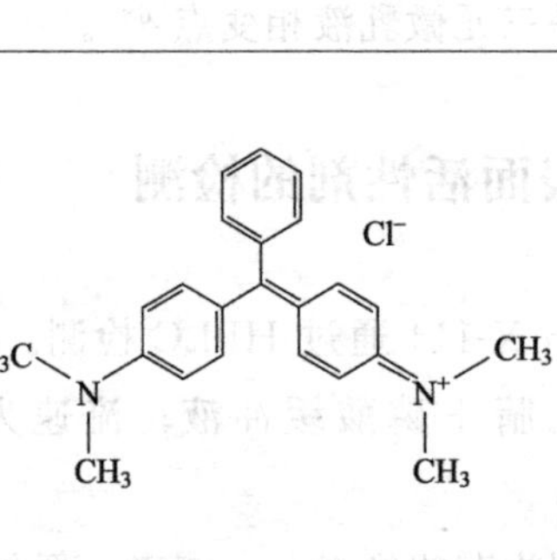	1.69	616

实验选用三种离子液体，分别为1-丁基-3-甲基咪唑四氟硼酸盐（1-butyl-3-methylimidazolium tetrafluoroborate，[Bmim]BF_4）、1-丁基-3-甲基咪唑六氟磷酸盐（1-butyl-3-methylimidazolium hexafluorophosphate，[Bmim]PF_6）和1-丁基-3-甲基咪唑氯盐（1-butyl-3-methylimidazolium chloride，[Bmim]Cl）。测定离子液体主要通过称重法。

6.2.4 建立标准曲线

在非离子表面活性剂Triton X-114及五种染料的检测限内，从低到高选择合适的浓度，以检测值对浓度作图得到标准曲线。后面实验中的物质含量由此标准曲线计算得到。

如图6-5～图6-10所示，分别代表Triton X-114、茜素、苋菜红、甲基橙、结晶紫和孔雀石绿的标准曲线。由于茜素具有电离性，在不同的pH下吸光度发生变化。因此，实验测定了不同pH下茜素的标准曲线。图6-6(a) 代表pH＜7时茜素的标准曲线，图6-6(b) 代表pH＝11时茜素的标准曲线，图6-6(c) 代表pH＝13时茜素的标准曲线。

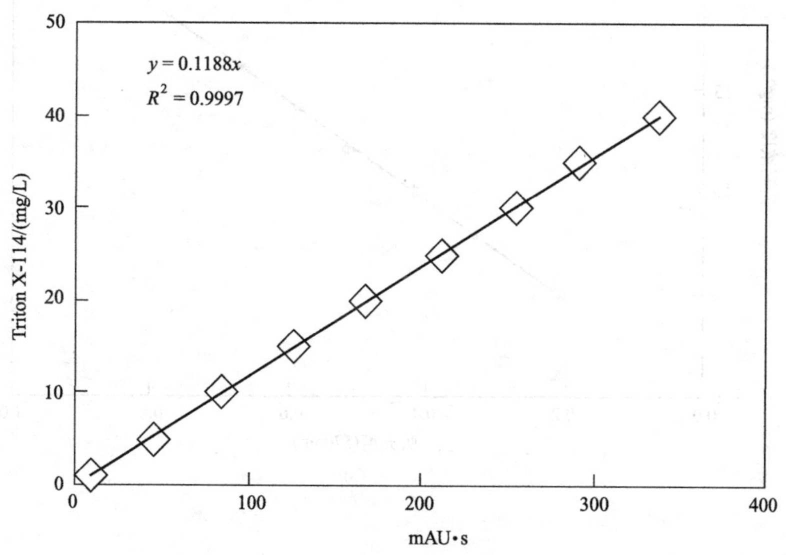

图6-5 Triton X-114的标准曲线

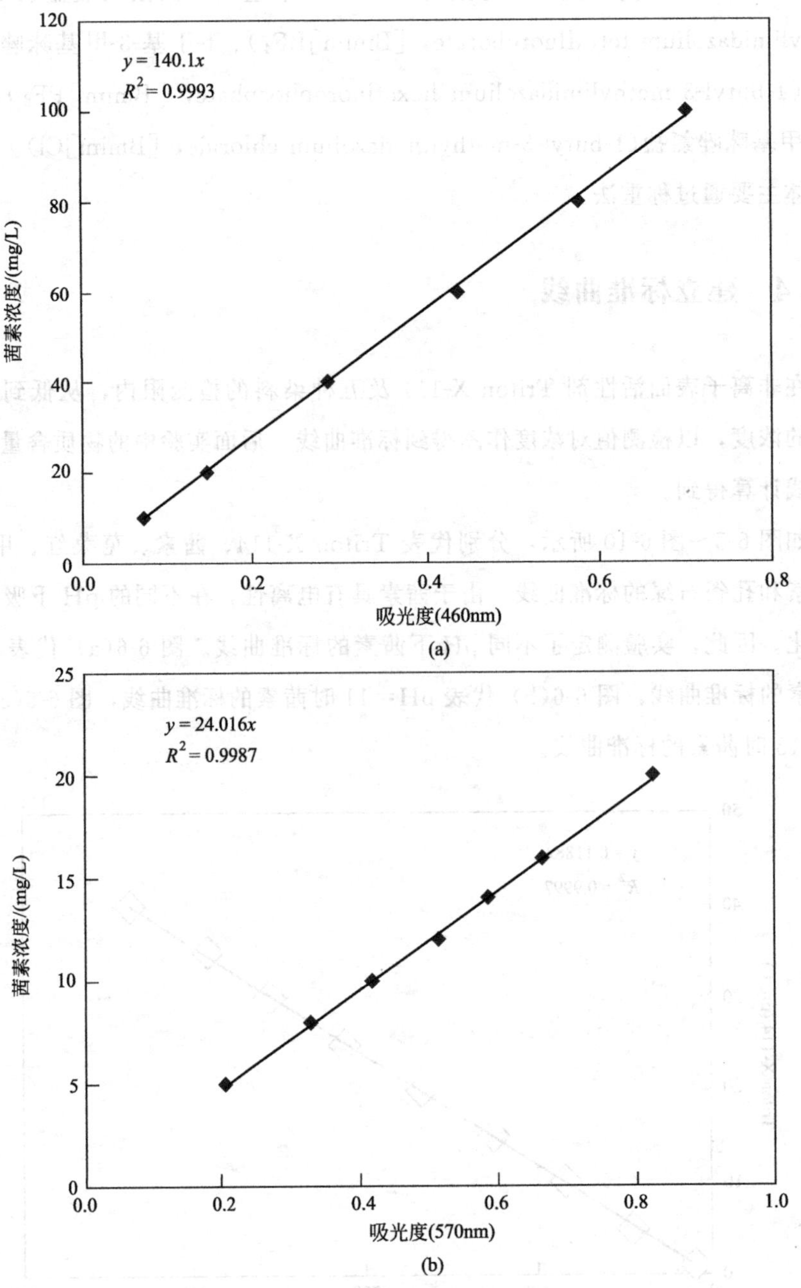
y = 140.1x
R² = 0.9993
茜素浓度/(mg/L)
吸光度(460nm)
(a)
y = 24.016x
R² = 0.9987
茜素浓度/(mg/L)
吸光度(570nm)
(b)

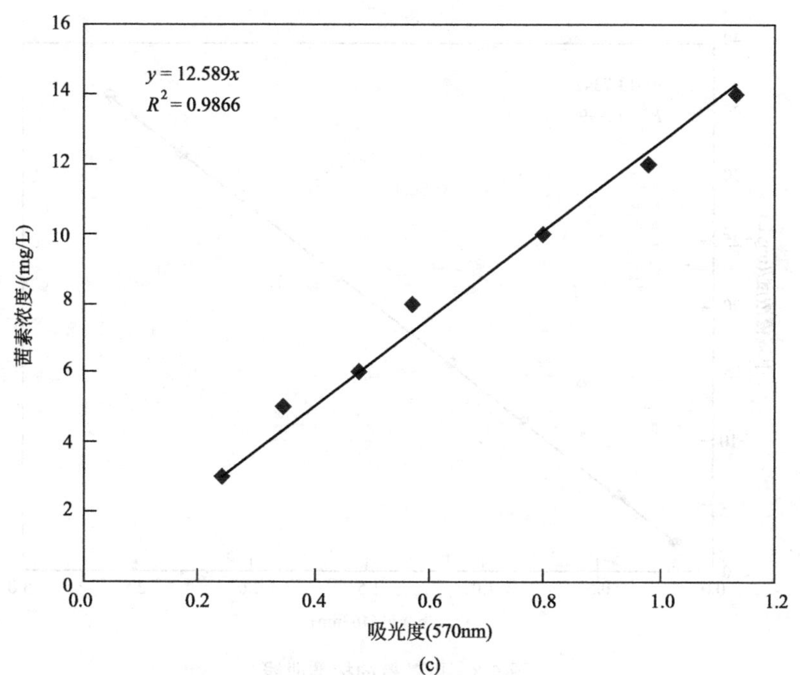

图 6-6 茜素的标准曲线

(a) pH<7；(b) pH＝11；(c) pH＝13

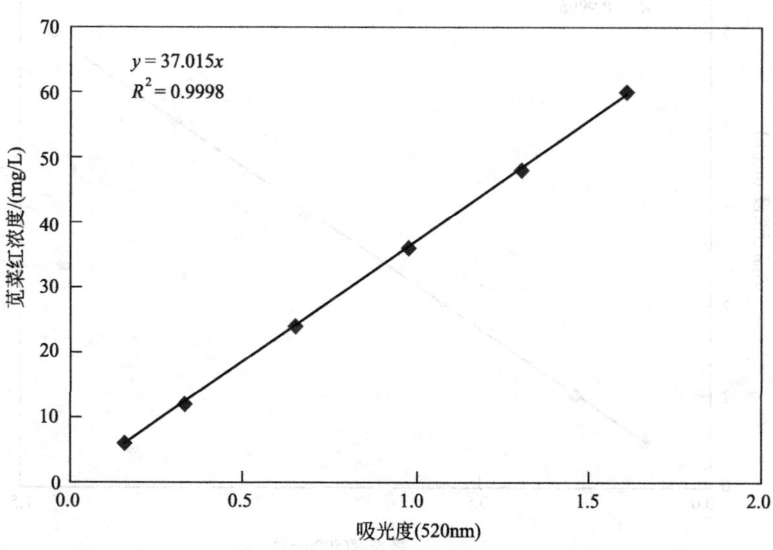

图 6-7 苋菜红的标准曲线

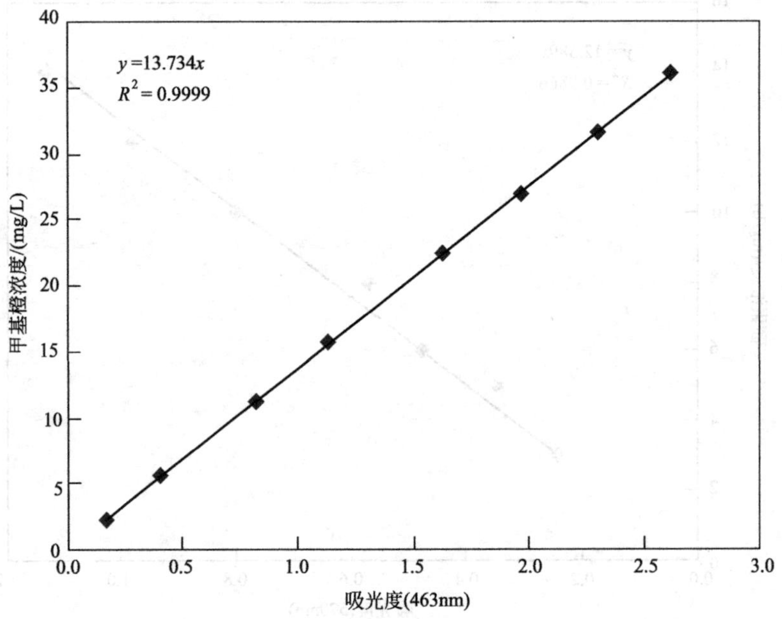

图 6-8　甲基橙的标准曲线

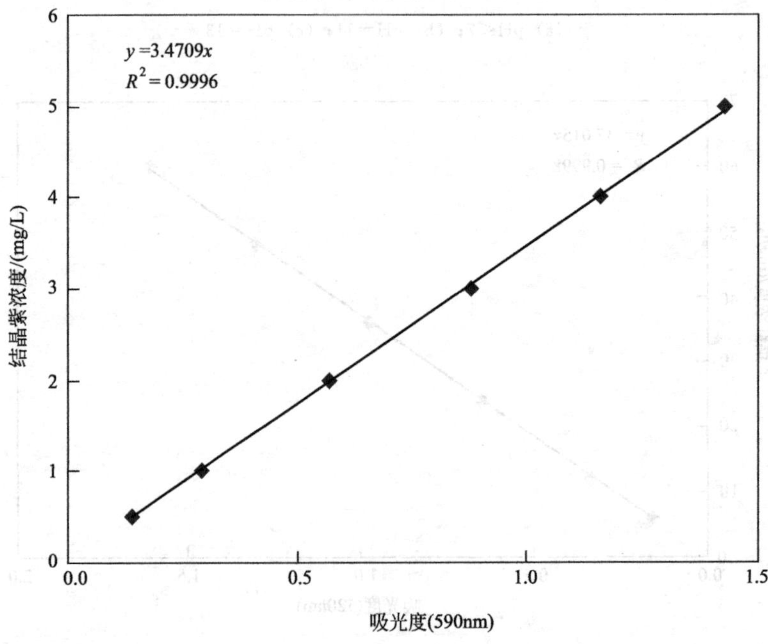

图 6-9　结晶紫的标准曲线

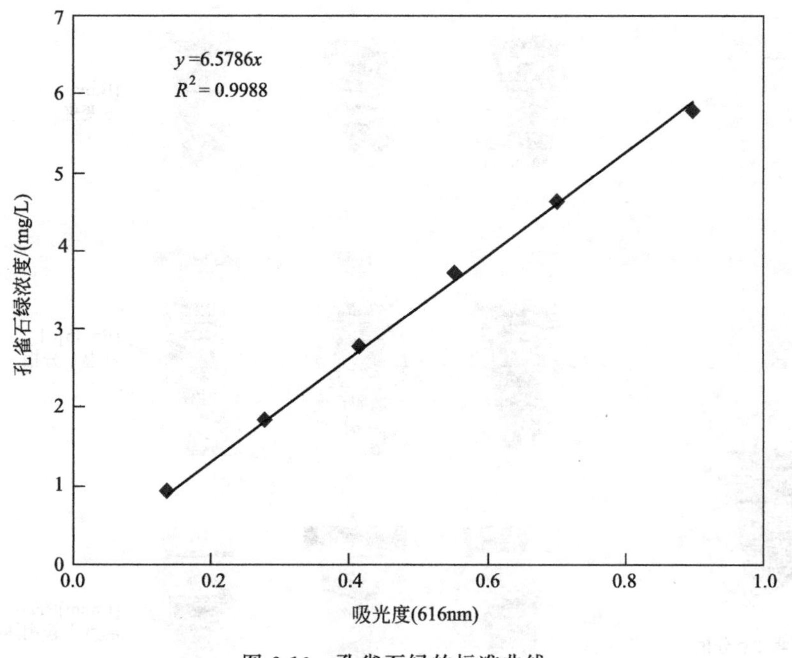

图 6-10　孔雀石绿的标准曲线

6.3 三元微乳液回收非离子表面活性剂的相分离机理

6.3.1 离子液体的筛选

本实验选择非离子表面活性剂 Triton X-114 形成的浊点系统作为研究对象。在染料结晶紫的浊点萃取后，取 1mL 含结晶紫的凝聚层相与等体积的离子液体混合，观察相变化。

如图 6-11 所示，室温（25℃）下，离子液体［Bmim］BF_4 和［Bmim］Cl 与 TX-114 的凝聚层相形成均匀的单相系统。高温状态下（85℃），［Bmim］BF_4 形成的单相系统没有发生变化。而［Bmim］Cl 形成的单相系统发生相分离，上相是水相，下相是离子液体相。结晶紫在两相间不均匀分配，但明显［Bmim］Cl

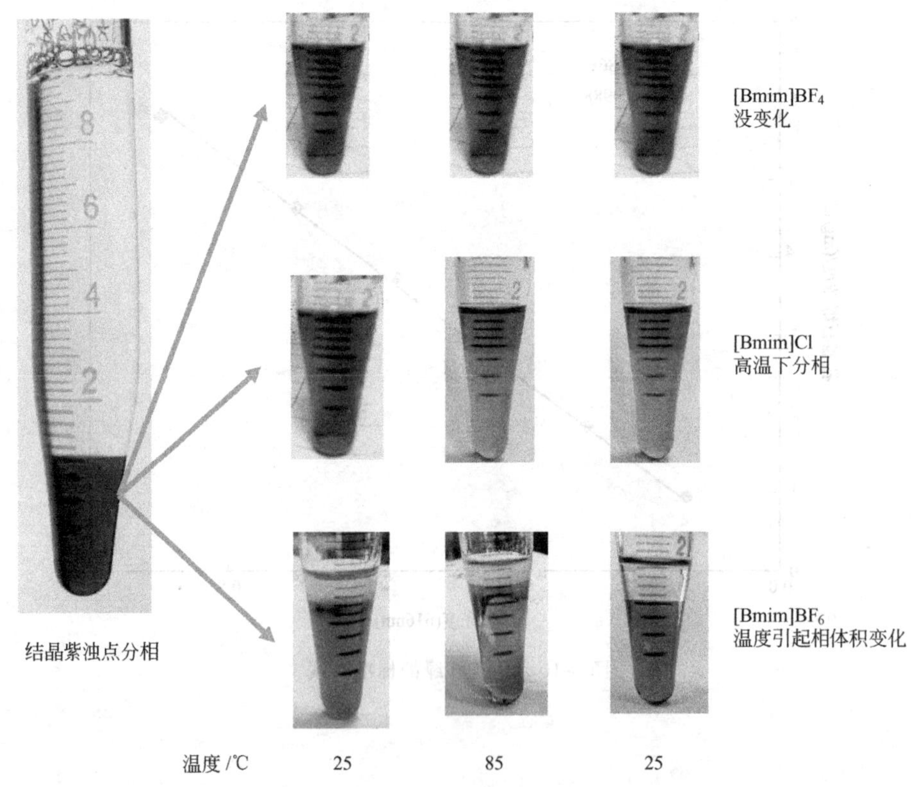

图 6-11 三种离子液体与浊点系统凝聚层相的相互作用

对结晶紫的萃取不彻底，更多的染料保留在水相。

而离子液体［Bmim］PF_6 与 TX-114 凝聚层相形成两相系统，上相呈乳状的不透明分散液，下相含有结晶紫，成紫色。初步分析，此两相系统上相为水相，下相为离子液体相，结晶紫从 TX-114 凝聚层相中被萃取到离子液体相中。发现室温下，下相离子液体相的体积接近 1.4mL，判断 TX-114 被萃取到了下相当中，85℃下恒温水浴后，体积比并未改变。室温下过夜平衡后，试管内两相体积发生了很大变化，上相体积增大，下相体积接近 1mL。同时，用枪头吸取上相澄清溶液一点滴在指头上，手指捻过有黏黏的感觉。由此可初步确定，表面活性剂已经重新分配在两相系统的水相当中。因此，利用离子液体分离浊点系统萃取染料后的凝聚层相中的表面活性剂和染料，初步判断可行。

6.3.2 表面活性剂-水-离子液体三元微乳液的建立

本实验采用直接观察法[44]测定了 Triton X-114＋水＋［Bmim］PF_6 的三元微

乳液体系的两相边界，在图 6-12 中用粗线条表示。左侧为两相区，右侧为单相区。

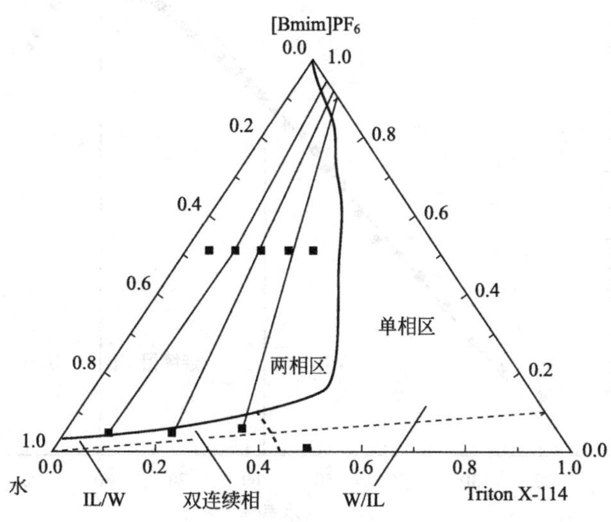

图 6-12　Triton X-114＋水＋[Bmim]PF_6 的三元相图

在单相区内，以点（水：Triton X-114：[Bmim]PF_6＝1：0：0）和（水：Triton X-114：[Bmim]PF_6＝0：0.9：0.1）之间的虚线作为研究路线，采用电导法[69]研究单相区内微乳液的组成情况。如图 6-13 所示，随着体系内含水率的逐渐升高，体系的电导率呈现规律性的变化。

Clausse 等[68]在 1981 年 *Nature* 上发表了关于电导率测定微乳液类型的详细方法。在图 6-13 中，三元体系的电导率在含水率为 57%和 80%时产生拐点，表示这种情况下溶液的微乳液类型发生了变化。

在离子液体包水（W/IL）微乳液中，随着含水率的增大，离子液体[Bmin]PF_6 水解量增加，即导电离子的浓度增大；同时水核量增加，水核与水核碰撞的概率增加，导致体系电导率增大。当含水率增加到一定程度，离子液体包容更多的水核，导致体系中出现自由水。自由水的出现，令体系的电导率进一步升高。但同时由于离子液体已经完全水解，总体积的增大导致电荷密度的降低，因此电导率的增加趋于缓慢。此时溶液处于双连续相（bicontinuous phase）。当含水率增大到一定程度，体系中离子液体被包裹起来，形成水化膜，溶液电导率开始下降。此时微乳液类型为水包离子液体（IL/W）。

根据电导率数据得到的相变水含量在图 6-12 中表示出来，可直观了解单相区的微乳液组成。了解三元体系的微乳液组成，在环境污染物治理、生物催化、手性合成、纳米材料合成等方面具有重要意义[69-73]。

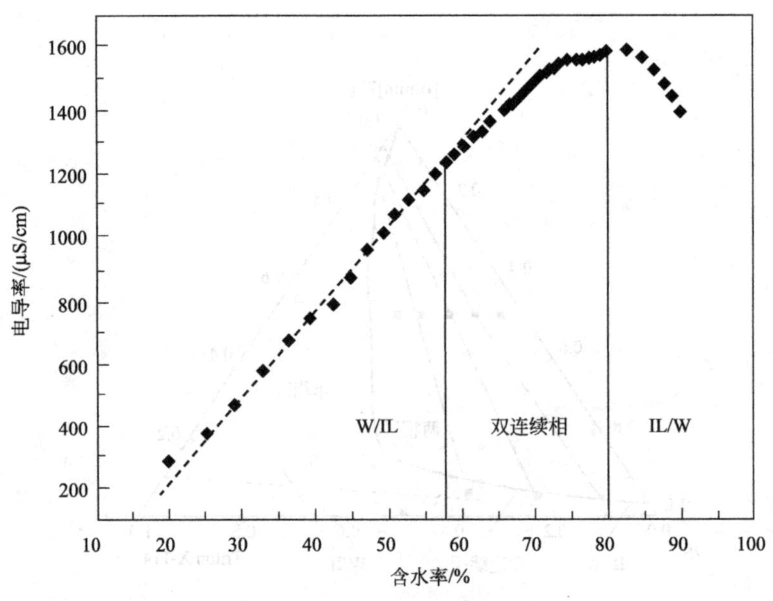

图 6-13　电导率测定微乳液类型

6.3.3　温度调节的偏析型相分离和缔合型相分离

在图 6-12 中，选取五个点研究 Triton X-114＋水＋[Bmim]PF_6 三元体系的相组成。不同温度下，三元体系的相组成发生变化，如图 6-14 所示。从图中可以明显看出，常温下，溶液分为上下两部分。上相是 Triton X-114 水溶液相，下相是离子液体 [Bmim]PF_6 相；加热后，上相体积明显减小，Triton X-114 进入下相离子液体相，系统从偏析型相分离转变为缔合型相分离。

对于非离子表面活性剂的回收，不同的相分离体系，可采用不同的策略。对于缔合型相分离，非离子表面活性剂和室温离子液体共处一相。此体系可用于高水溶性物质的分离，从浊点系统中萃取非离子表面活性剂而留溶质在水相。对于偏析型相分离，非离子表面活性剂和室温离子液体分别在上下两相。应用此体系，可从浊点系统凝聚层相中萃取溶质而把非离子表面活性剂留在水相。在本实验中，从前文可以看出，染料结晶紫主要分配在离子液体相。因此，偏析型相分离适合作为本实验非离子表面活性剂的回收体系。

在常温下分相后，Triton X-114＋水＋[Bmim]PF_6 形成偏析型相分离系统。溶液相组成情况如表 6-4 所示。

图 6-14 不同温度下 Triton X-114＋水＋[Bmim]PF_6 三元体系的相分离

表 6-4 偏析型相分离系统组分分析

初始值/g			分相后上相/g			分相后下相/g		
Triton X-114	H_2O	[Bmim]PF_6	Triton X-114	H_2O	[Bmim]PF_6	Triton X-114	H_2O	[Bmim]PF_6
0.10	0.90	1.00	0.002	0.99	0.01	0.10	−0.09	0.99
0.20	0.80	1.00	0.08	0.88	0.04	0.12	−0.08	0.96
0.30	0.70	1.00	0.20	0.75	0.05	0.10	−0.05	0.95
0.40	0.60	1.00	0.33	0.61	0.06	0.07	−0.01	0.94
0.50	0.50	1.00	0.49	0.51	0.00	0.01	−0.01	1.00

表 6-4 中上下相组成在图 6-12 中表示，连接上下相组成的线经过原始组成点，此线称为系线，两相组成点叫作临界点，基本与两相分界线吻合。

从表 6-4 中数据分析可知，[Bmim]PF_6 基本分配在下相，水则保留在上相。这主要和 [Bmim]PF_6 的基本性质有关。[Bmim]PF_6 是一类高疏水性室温离子液体，几乎不溶于水。在 Triton X-114＋水＋[Bmim]PF_6 三相体系中，Triton X-114 对 [Bmim]PF_6 在水中的溶解度影响不大。从数据分析可知，低浓度下，Triton X-114 主要分配在下相。高浓度下，Triton X-114 主要分配在水相。在浊点系统中，凝聚层相的非离子表面活性剂浓度较高。因此，此实验结果表明用离子液体 [Bmim]PF_6 回收非离子表面活性剂 Triton X-114 具有可行性。

6.3.4 偏析型相分离回收非离子表面活性剂

在一定温度下，五种染料的 Triton X-114 浊点萃取后，凝聚层相体积约为 1mL。小心移除稀相后，在 1mL 的凝聚层相里加入 1g [Bmim]PF_6（约 0.732mL），混匀后室温（25℃）下分相，结果如图 6-15 所示。

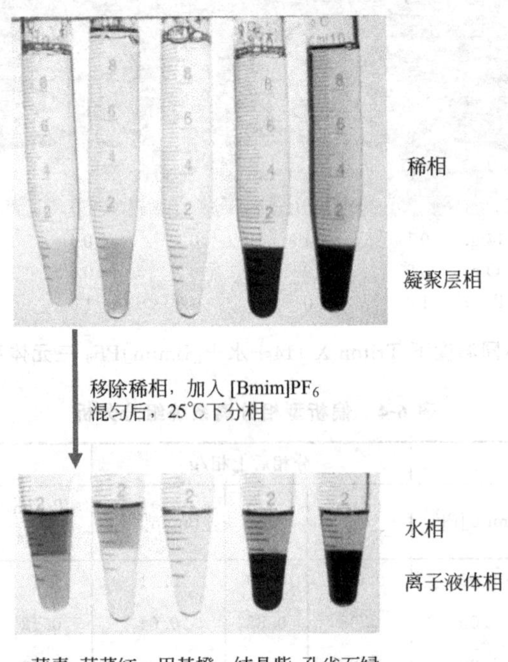

图 6-15 [Bmim]PF_6 分离染料回收 Triton X-114

记录相体积并分析水相中染料、Triton X-114 和 [Bmim]PF_6 浓度。计算染料在两相间的分配率和 Triton X-114 的回收率（表 6-5）。

对于苋菜红和甲基橙两种偶氮染料，离子液体无法分离染料回收非离子表面活性剂。这可能和偶氮染料特有的基团有关，而与其疏水性无关（苋菜红的 lg P 为 −1.98，甲基橙的 lg P 为 3.47）。传统溶剂，如有机溶剂、聚合物和非离子表面活性剂等，是分子溶剂，而离子液体是离子溶剂，这是离子液体区别于常规溶剂的典型结构特征[2]。这一特征导致离子液体相对于传统分子溶剂具有一些独特的性质。离子液体一方面表现出非极性有机溶剂的特征，同时具有离子性而表现出强极性有机溶剂的特征[74]。因此，离子液体 [Bmim]PF_6 对染料的萃取并非像有机溶剂或者浊点萃取那样与 lg P 值关联密切。其独特的分子性质令

表 6-5 偏析型相分离体系中的染料分配率及 Triton X-114 回收率

染料	染料分配率/%①	Triton X-114 回收率/%②
茜素	1.47	76.80±1.08
苋莱红	0	—③
甲基橙	0.29	—
结晶紫	23.87	83.11±0.37
孔雀石绿	38.55	96.58±1.70

① 染料分配率=(离子液体相中染料浓度)/(水相中染料浓度)；

② Triton X-114 回收率=(水相中 Triton X-114 浓度)/(凝聚层相中 Triton X-114 浓度)×100%；

③ 两种偶氮染料，苋莱红与甲基橙，主要分配在水相，说明通过离子液体分离染料回收非离子表面活性剂的方法不适合偶氮染料。

其和不同分子的相互作用体现出不同的机理。

[Bmim]PF_6 从浊点系统凝聚层相中分离出茜素、结晶紫和孔雀石绿，进而回收了非离子表面活性剂 Triton X-114。Triton X-114 的回收率基本保持在 80% 以上。这一分离过程不需要高温高压等高耗能设备，同时由于离子液体和非离子表面活性剂不具有挥发性，整个过程环境友好、无污染，具有巨大的应用潜力。

6.3.5 离子液体的回收

回收 Triton X-114 以后，染料分配在 [Bmim]PF_6 离子液体相中。需要进一步回收 [Bmim]PF_6 以节约成本，减少试剂浪费。从离子液体中回收化合物的方法有很多，包括水反萃、有机溶剂萃取、超临界二氧化碳萃取、pH 调节，甚至直接光降解[64,67,75-81]。

对于三苯基甲烷染料结晶紫和孔雀石绿，在尝试了水反萃、pH 调节和溶剂萃取（包括乙醇、丙醇、二氯甲烷、氯仿、乙酸乙酯等）后，也没能将染料从 [Bmim]PF_6 中分离出来。其原因可能与离子液体 [Bmim]PF_6 对三苯基甲烷染料结晶紫和孔雀石绿的萃取机理有关。Li 等发现，在用离子液体萃取染料时，若其萃取机理是离子交换，那么从离子液体中回收染料时比较困难[82]。因为通过离子交换，离子液体的自身结构受到破坏。超临界二氧化碳萃取技术在离子液体的回收中应用最广泛[76,79]。由于超临界二氧化碳基本溶解离子液体，实现了很多苯系物在离子液体中的反萃，研究较为透彻[76]。而原位光降解技术可以不分离化合物与离子液体，直接将离子液体中的化合物降解掉，在环境污染物去除方面具有良好的应用前景[81]。但由于实验条件限制，这两种方法在本实验中没有尝试。

通过 pH 调节的方法可以有效地从离子液体相中分离出蒽醌染料茜素，结果如图 6-16 所示。用离子液体［Bmim］PF_6 分离茜素回收非离子表面活性剂 Triton X-114 之后，移除含有 Triton X-114 的水相，在含有染料茜素的离子液体相中添加 1mL 不同 pH 值的水。混匀后室温静置分离，结果表明：酸性条件下（pH＜7），茜素主要分配在离子液体相；碱性条件下（pH＞7），茜素被反萃到水相，实现了离子液体［Bmim］PF_6 的回收。

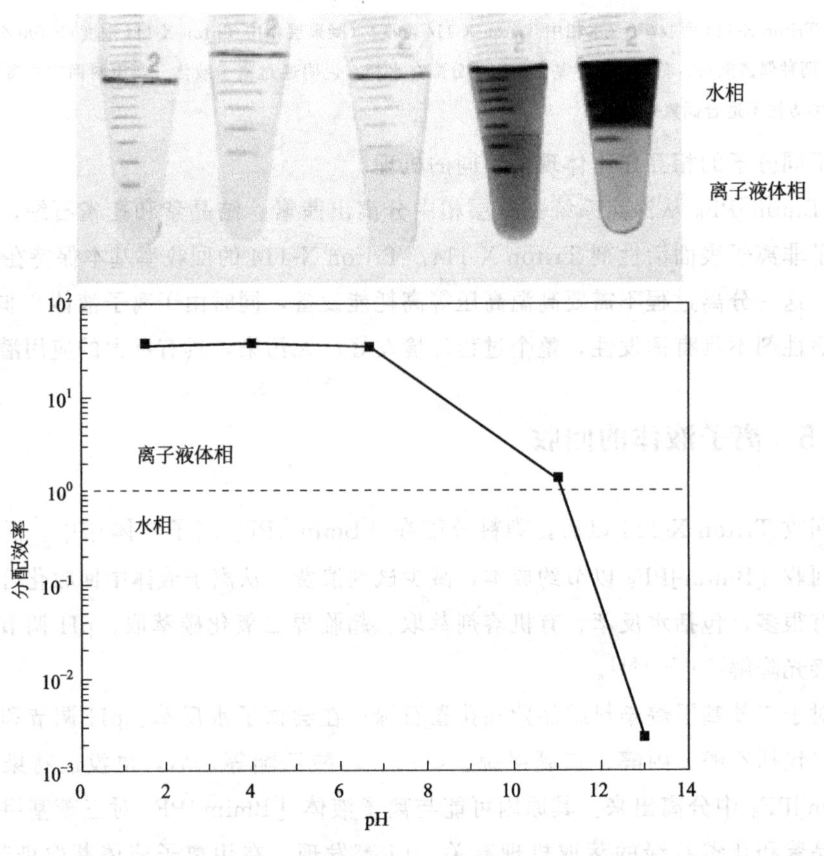

图 6-16　pH 调节分离染料回收［Bmim］PF_6

离子化合物在离子液体-水两相体系中的分配系数受到溶液 pH 的强烈影响[2]。茜素在酸性条件下呈电中性，主要分配在下相离子液体相。在碱性条件下，茜素发生电离而主要分配到水相。同样，其他有机酸、碱在水-［Bmim］PF_6 两相体系中也有相类似的分配规律[65,80]。

6.3.6 离子液体回收混合非离子表面活性剂

在上面的实验中，通过离子液体 [Bmim]PF_6 实现了非离子表面活性剂 Triton X-114 的回收。本节尝试离子液体 [Bmim]PF_6 用于分离染料回收混合非离子表面活性剂 Brij 30＋TMN-3。

(1) 温度对相行为的影响

实验考察了不同温度下水-[Bmim]PF_6-混合非离子表面活性剂体系的相行为，如图 6-17 所示。根据相行为结果考虑不同的染料分离和表面活性剂回收策略。

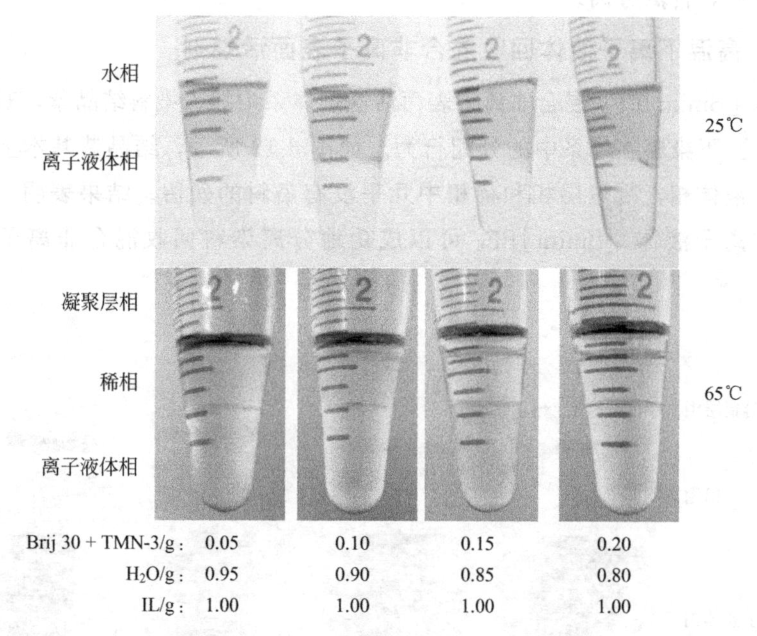

图 6-17 温度对水-[Bmim]PF_6-混合非离子表面活性剂体系相行为的影响

常温（25℃）下，将混合非离子表面活性剂水溶液与离子液体 [Bmim]PF_6 混合时，体系的黏度非常大，呈凝胶状。混匀后，经过 2000r/min 离心 30min 才能将体系分离彻底。如图 6-17 所示，常温下水-[Bmim]PF_6-混合非离子表面活性剂体系分为上下两相。上相为水相，为混合非离子表面活性剂水溶液；下相为离子液体相。从体积上判断，离子液体相没有溶解非离子表面活性剂。加热到 65℃以后，溶液黏度大大降低，无需离心即可彻底分相。溶液分为上中下三相。高温下，水相的混合非离子表面活性剂水溶液发生浊点相分离，上相为凝聚层相，下相为稀相。而离子液体相没有发生变化。

前文提到，Triton X-114＋水＋[Bmim]PF_6 三相体系在常温下分为两相，这与水-[Bmim]PF_6-混合非离子表面活性剂在常温下的相行为相似。但在高温下，Triton X-114＋水＋[Bmim]PF_6 三相体系中，非离子表面活性剂 Triton X-114 被萃取到离子液体相。而在水-[Bmim]PF_6-混合非离子表面活性剂体系中，混合非离子表面活性剂在高温下依然保留在水相。

由于无论常温还是高温下，混合非离子表面活性剂 Brij 30＋TMN-3 都没有被萃取到离子液体相。这为离子液体 [Bmim]PF_6 分离染料回收混合非离子表面活性剂 Brij 30＋TMN-3 提供了新的思路。策略一：常温下，染料分配在离子液体相，借助离心作用完成相分离。策略二：高温下，溶液形成三相，染料分配在离子液体相，自然分离。

(2) 高温下离子液体回收混合非离子表面活性剂

向水-[Bmim]PF_6-混合非离子表面活性剂体系中加入染料结晶紫，观察在高温（65℃）下染料在体系中的分配行为。如图 6-18 所示，结晶紫基本上全部萃取到离子液体相，凝聚层相和稀相中几乎没有染料的残留。结果表明，在高温下，利用离子液体 [Bmim]PF_6 可以成功地分离染料回收混合非离子表面活性剂。

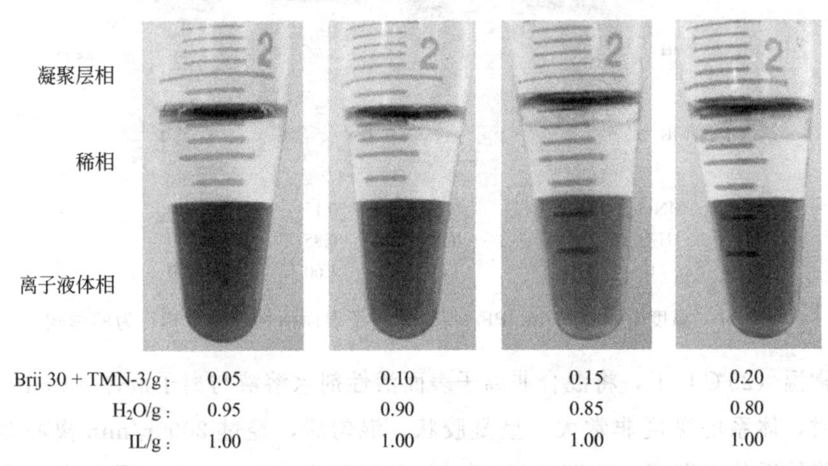

图 6-18 [Bmim]PF_6 在 65℃下回收 Brij 30＋TMN-3

(3) 常温下离子液体回收混合非离子表面活性剂

在结晶紫的萃取微生物脱色之后，凝聚层相中的染料用 [Bmim]PF_6 分离以回收混合非离子表面活性剂 Brij 30＋TMN-3。如图 6-19 所示，在常温（25℃）

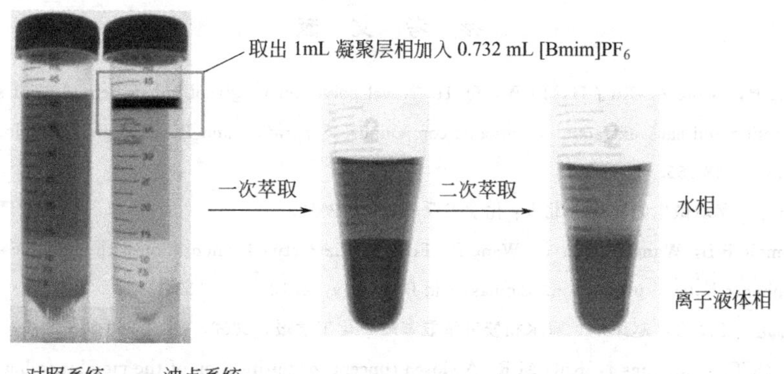

图 6-19 [Bmim]PF_6 用于结晶紫萃取微生物脱色后非离子表面活性剂的回收

下，结晶紫在 Brij 30＋TMN-3 混合非离子表面活性剂水溶液形成的浊点系统中经菌种嗜水气单胞菌 DN322p 萃取微生物脱色后，残留的结晶紫和隐性结晶紫被萃取到凝聚层相中。取出凝聚层相，加入等质量的离子液体［Bmim］PF_6 混匀后，常温下离心分相。分相后，染料主要被萃取到离子液体相中。一次萃取后，水相中除了非离子表面活性剂之外，还有很多染料。经二次萃取后，水相中染料浓度明显降低。结果说明，通过多级萃取可以彻底将凝聚层相中的染料萃取出来，进而回收混合非离子表面活性剂 Brij 30＋TMN-3。

浊点系统作为一种绿色体系可代替传统的两相系统应用在两相催化、萃取微生物转化和萃取发酵、萃取及分离过程等多个多领域[42,56-60]。然而，关于从浊点系统中分离出有机物并回收表面活性剂方法的报道不多[1,19]，这严重限制了浊点系统中工业化应用。

本实验详细分析了 Triton X-114＋水＋［Bmim］PF_6 三元体系的相组成，微乳液结构及温度诱导的偏析型相分离和缔合型相分离。根据两种不同的相分离类型，假设出两种可行的有机物分离及非离子表面活性剂回收策略。

创新地应用偏析型相分离方法从浊点系统凝聚层相中分离出染料并回收了非离子表面活性剂 Triton X-114。同时，探讨了离子液体的回收方法。通过 pH 调节的方式控制可电离染料在水-离子液体两相体系中的分配系数，达到回收离子液体的目的。

用离子液体回收非离子表面活性剂的方法，避免了有机溶剂的介入。操作过程绿色无污染，效率高，具有广阔的工业化应用前景。但如何回收离子液体，技术还不成熟，需要继续探索。

参 考 文 献

[1] Liang R，Wang Z，Xu J H，Li W，Qi H. Novel polyethylene glycol induced cloud point system for extraction and back-extraction of organic compounds. Separation and Purification Technology，2009，66 (2)：248-256.

[2] 王志龙．萃取微生物转化．北京：化学工业出版社，2012.

[3] Dhamole P B，Wang Z，Liu Y，Wang B，Feng H. Extractive fermentation with non-ionic surfactants to enhance butanol production. Biomass and Bioenergy，2012，40：112-119.

[4] 王志龙，王劲松，赵凤生．双水相胶束萃取苯酚．化工学报，2002，(3)：269-273.

[5] Minuth T，Thömmes J，Kula M R. A closed concept for purification of the membrane-bound cholesterol oxidase from *Nocardia rhodochrous* by surfactant-based cloud-point extraction，organic-solvent extraction and anion-exchange chromatography. Biotechnology And Applied Biochemistry，1996，23 (2)：107-116.

[6] Collén A，Persson J，Linder M，Nakari-Setälä T，Penttilä M，Tjerneld F，Sivars U. A novel two-step extraction method with detergent/polymer systems for primary recovery of the fusion protein endoglucanase I-hydrophobin I. Biochimica et Biophysica Acta—General Subjects，2002，1569 (1-3)：139-150.

[7] Draye M，Thomas S，Cote G，Favre-Reguillon A，LeBuzit G，Guy A，Foos J. Cloud-point extraction for selective removal of Gd(Ⅲ) and La(Ⅲ) with 8-hydroxyquinoline. Separation Science and Technology，2005，40 (1-3)：611-622.

[8] Sun C，Xie Y，Tian Q，Liu H. Cloud point extraction of glycyrrhizic acid from licorice root. Separation Science and Technology，2007，42 (14)：3259-3270.

[9] Cheng H，Sabatini D A. Separation of organic compounds from surfactant solutions：A review. Separation Science and Technology，2007，42 (3)：453-475.

[10] Quina F H，Hinze W L. Surfactant-mediated cloud point extractions：An environmentally benign alternative separation approach. Industrial and Engineering Chemistry Research，1999，38 (11)：4150-4168.

[11] Ibrahim N M，Wheals B B. Oligomeric separation of alkylphenol ethoxylate surfactants on silica using aqueous acetonitrile eluents. Journal of Chromatography A，1996，731 (1-2)：171-177.

[12] Willauer H D，Huddleston J G，Rogers R D. Solute partitioning in aqueous biphasic systems composed of polyethylene glycol and salt：the partitioning of small neutral organic species. Industrial and engineering chemistry research，2002，41 (7)：1892-1904.

[13] Frankewich R P，Hinze W L. Evaluation and optimization of the factors affecting nonionic surfactant-mediated phase separations. Analytical Chemistry，1994，66 (7)：944-954.

[14] Jiang J S，Vane L M，Sikdar S K. Recovery of VOCs from surfactant solutions by pervaporation. Journal of Membrane Science，1997，136 (1-2)：233-247.

[15] Collén A，Persson J，Linder M，Nakari-Setälä T，Penttilä M，Tjerneld F，Sivars U. A novel two-step extraction method with detergent/polymer systems for primary recovery of the fusion protein en-

doglucanase I-hydrophobin I. Biochimica et Biophysica Acta (BBA)—General Subjects, 2002, 1569 (1-3): 139-150.

[16] Olsson U, Shinoda K, Lindman B. Change of the structure of microemulsions with the hydrophile-lipophile balance of nonionic surfactant as revealed by NMR self-diffusion studies. Journal Of Physical Chemistry, 1986, 90 (17): 4083-4088.

[17] Wang Z, Xu J H, Liang R, Qi H. A downstream process with microemulsion extraction for microbial transformation in cloud point system. Biochemical Engineering Journal, 2008, 41 (1): 24-29.

[18] Tricoli V, Farnesi M, Nicolella C. Bicontinuous microemulsions as adsorbents for liquid-phase separation/purification. Aiche Journal, 2006, 52 (8): 2767-2773.

[19] Pan T, Wang Z, Xu J H, Wu Z, Qi H. Stripping of nonionic surfactants from the coacervate phase of cloud point system for lipase separation by Winsor II microemulsion extraction with the direct addition of alcohols. Process Biochemistry, 2010, 45 (5): 771-776.

[20] 赵伟，沈安，江宁．微生物转化生产 L-苯基乙酰基甲醇（L-PAC）研究进展．应用与环境生物学报，2003，(2)：218-220.

[21] Long A, Ward O P. Biotransformation of benzaldehyde by *Saccharomyces cerevisiae*: characterization of the fermentation and toxicity effects of substrates and products. Biotechnology and bioengineering, 1989, 34 (7): 933-941.

[22] Long A, Ward O P. Biotransformation of aromatic aldehydes by *Saccharomyces cerevisiae*: investigation of reaction rates. Journal of industrial microbiology, 1989, 4 (1): 49-53.

[23] Kujawska A, Kujawski J, Bryjak M, Kujawski W. ABE fermentation products recovery methods—a review. Renewable and Sustainable Energy Reviews, 2015, 48: 648-661.

[24] Qureshi N, Saha B C, Cotta M A, Singh V. An economic evaluation of biological conversion of wheat straw to butanol: a biofuel. Energy Conversion and Management, 2013, 65: 456-462.

[25] Rakopoulos D C, Rakopoulos C D, Giakoumis E G, Dimaratos A M, Kyritsis D C. Effects of butanol-diesel fuel blends on the performance and emissions of a high-speed DI diesel engine. Energy Conversion and Management, 2010, 51 (10): 1989-1997.

[26] Qureshi N, Hughes S, Maddox I S, Cotta M A. Energy-efficient recovery of butanol from model solutions and fermentation broth by adsorption. Bioprocess and biosystems engineering, 2005, 27 (4): 215-222.

[27] Levario T J, Dai M, Yuan W, Vogt B D, Nielsen D R. Rapid adsorption of alcohol biofuels by high surface area mesoporous carbons. Microporous and Mesoporous Materials, 2012, 148 (1): 107-114.

[28] Oudshoorn A, Van Der Wielen L A, Straathof A J. Assessment of options for selective 1-butanol recovery from aqueous solution. Industrial and Engineering Chemistry Research, 2009, 48 (15): 7325-7336.

[29] Qureshi N, Blaschek H P. Production of acetone butanol ethanol (ABE) by a hyper-producing mutant strain of *Clostridium beijerinckii* BA101 and recovery by pervaporation. Biotechnology progress, 1999, 15 (4): 594-602.

[30] Rozicka A, Niemistö J, Keiski R L, Kujawski W. Apparent and intrinsic properties of commercial PDMS based membranes in pervaporative removal of acetone, butanol and ethanol from binary aqueous mixtures. Journal of Membrane Science, 2014, 453: 108-118.

[31] Li S Y, Srivastava R, Parnas R S. Separation of 1-butanol by pervaporation using a novel tri-layer PDMS composite membrane. Journal of Membrane Science, 2010, 363 (1-2): 287-294.

[32] Qureshi N, Maddox I S. Reduction in butanol inhibition by perstraction: utilization of concentrated lactose/whey permeate by *Clostridium acetobutylicum* to enhance butanol fermentation economics. Food and Bioproducts Processing, 2005, 83 (1): 43-52.

[33] Evans P J, Wang H Y. Enhancement of butanol formation by Clostridium acetobutylicum in the presence of decanol-oleyl alcohol mixed extractants. Applied and Environmental Microbiology, 1988, 54 (7): 1662-1667.

[34] Taconi K A, Venkataramanan K P, Johnson D T. Growth and solvent production by *Clostridium pasteurianum* ATCC® 6013™ utilizing biodiesel-derived crude glycerol as the sole carbon source. Environmental Progress and Sustainable Energy: An Official Publication of the American Institute of Chemical Engineers, 2009, 28 (1): 100-110.

[35] Qureshi N, Blaschek H P. Recovery of butanol from fermentation broth by gas stripping. Renewable Energy, 2001, 22 (4): 557-564.

[36] Ezeji T C, Karcher P M, Qureshi N, Blaschek H P. Improving performance of a gas stripping-based recovery system to remove butanol from *Clostridium beijerinckii* fermentation. Bioprocess and biosystems engineering, 2005, 27 (3): 207-214.

[37] Dhamole P B, Mane R G, Feng H. Screening of non-ionic surfactant for enhancing biobutanol production. Applied biochemistry and biotechnology, 2015, 177 (6): 1272-1281.

[38] Singh K, Gedam P S, Raut A N, Dhamole P B, Dhakephalkar P K, Ranade D R. Enhanced *n*-butanol production by *Clostridium beijerinckii* MCMB 581 in presence of selected surfactant. 3 Biotech, 2017, 7 (3): 161.

[39] Raut A N, Gedam P S, Dhamole P B. Back-extraction of butanol from coacervate phase using Winsor Ⅲ microemulsion. Process Biochemistry, 2018, 70: 160-167.

[40] Melani N B, Tambourgi E B, Silveira E. Lipases: from production to applications. Separation and Purification Reviews, 2020, 49 (2): 143-158.

[41] Zhao L L, Xu J H, Zhao J, Pan J, Wang Z L. Biochemical properties and potential applications of an organic solvent-tolerant lipase isolated from *Serratia marcescens* ECU1010. Process Biochemistry, 2008, 43 (6): 626-633.

[42] Pan T, Wang Z L, Xu J H, Wu Z Q, Qi H S. Extractive fermentation in cloud point system for lipase production by *Serratia marcescens* ECU1010. Appllied Microbioliology Biotechnology, 2010, 85 (6): 1789-1796.

[43] Ooi C W, Tan C P, Hii S L, Ariff A, Ibrahim S, Ling T C. Primary recovery of lipase derived from *Burkholderia* sp. ST8 with aqueous micellar two-phase system. Process biochemistry, 2011, 46 (9): 1847-1852.

[44] Gao Y, Han S, Han B, Li G, Shen D, Li Z, Du J, Hou W, Zhang G. TX-100/water/1-butyl-3-methylimidazolium hexafluorophosphate microemulsions. Langmuir, 2005, 21 (13): 5681-5684.

[45] Aki S N, Brennecke J F, Samanta A. How polar are room-temperature ionic liquids? Chemical Communications, 2001, (5): 413-414.

[46] Greaves T L, Drummond C J. Solvent nanostructure, the solvophobic effect and amphiphile self-assembly in ionic liquids. Chemical Society Reviews, 2013, 42 (3): 1096-1120.

[47] Inoue T, Misono T. Cloud point phenomena for POE-type nonionic surfactants in a model room temperature ionic liquid. Journal of colloid and interface science, 2008, 326 (2): 483-489.

[48] Inoue T, Misono T. Cloud point phenomena for POE-type nonionic surfactants in imidazolium-based ionic liquids: Effect of anion species of ionic liquids on the cloud point. Journal of colloid and interface science, 2009, 337 (1): 247-253.

[49] Inoue T, Iwasaki Y. Cloud point phenomena of polyoxyethylene-type surfactants in ionic liquid mixtures of emimBF4 and hmimBF4. Journal of colloid and interface science, 2010, 348 (2): 522-528.

[50] Álvarez M S, Rivas M, Deive F J, Sanromán M A, Rodríguez A. Ionic liquids and non-ionic surfactants: a new marriage for aqueous segregation. RSC advances, 2014, 4 (62): 32698-32700.

[51] Álvarez M S, Patiño F, Deive F J, Sanromán M Á, Rodríguez A. Aqueous immiscibility of cholinium chloride ionic liquid and Triton surfactants. The Journal of Chemical Thermodynamics, 2015, 91: 86-93.

[52] Álvarez M S, Esperança J M, Deive F J, Sanromán M Á, Rodríguez A. A biocompatible stepping stone for the removal of emerging contaminants. Separation and Purification Technology, 2015, 153: 91-98.

[53] He Y, Lodge T P. The micellar shuttle: Thermoreversible, intact transfer of block copolymer micelles between an ionic liquid and water. Journal of The American Chemical Society, 2006, 128 (39): 12666-12667.

[54] Zhao L, Zhang X, Wang Z. Extraction of anionic dyes with ionic liquid-nonionic surfactant aqueous two-phase system. Separation Science and Technology, 2017, 52 (5): 804-811.

[55] Cláudio A F M, Neves M C, Shimizu K, Lopes J N C, Freire M G, Coutinho J A. The magic of aqueous solutions of ionic liquids: ionic liquids as a powerful class of catanionic hydrotropes. Green Chemistry, 2015, 17 (7): 3948-3963.

[56] Wang Z. Extractive whole cell biotransformation in cloud point system // Wendt P L, Hoysted D S. Non-Ionic Surfactants. New York: Nova Science Publishers, 2010: 83-136.

[57] Wang Z, Dai Z. Extractive microbial fermentation in cloud point system. Enzyme and Microbial Technology, 2010, 46 (6): 407-418.

[58] Hinze W L, Pramauro E. A critical review of surfactant-mediated phase separations (cloud-point extractions): theory and applications. Critical Reviews in Analytical Chemistry, 1993, 24 (2): 133-177.

[59] Wang Z, Guo Y, Bao D, Qi H. Direct extraction of phenylacetic acid from immobilised enzymatic hydrolysis of penicillin G with cloud point extraction. Journal of Chemical Technology and Biotechnol-

ogy, 2006, 81 (4): 560-565.

[60] Wang Z. The potential of cloud point system as a novel two-phase partitioning system for biotransformation. Applied Microbiology and Biotechnology, 2007, 75 (1): 1-10.

[61] Terstappen G C, Geerts A J, Kula M R. The use of detergent-based aqueous two-phase systems for the isolation of extracellular proteins: purification of a lipase from Pseudomonas cepacia. Biotechnology and Applied Biochemistry, 1992, 16 (3): 228-235.

[62] Terstappen G C, Kula M R. Selective extraction and quantitation of polyoxyethylene detergents and its application in protein determination. Analytical Letters, 1990, 23 (12): 2175-2193.

[63] Bai Z, He Y, Young N P, Lodge T P. A thermoreversible micellization - Transfer - Demicellization shuttle between water and an ionic liquid. Macromolecules, 2008, 41 (18): 6615-6617.

[64] Pei Y C, Wang J J, Xuan X P, Fan J, Fan M H. Factors affecting ionic liquids based removal of anionic dyes from water. Environmental Science and Technology, 2007, 41 (14): 5090-5095.

[65] Huddleston J G, Willauer H D, Swatloski R P, Visser A E, Rogers R D. Room temperature ionic liquids as novel media for "clean" liquid-liquid extraction. Chemical Communications, 1998, (16): 1765-1766.

[66] Huang M H, Ma X G, Huang B. Application of ionic liquids to separation and analysis of environmental pollutants. Physical Testing and Chemical Analysis Part B: Chemical Analysis, 2007, 43 (11): 981-985.

[67] Vijayaraghavan R, Vedaraman N, Surianarayanan M, MacFarlane D R. Extraction and recovery of azo dyes into an ionic liquid. Talanta, 2006, 69 (5): 1059-1062.

[68] Clausse M, Peyrelasse J, Heil J, Boned C, Lagourette B. Bicontinuous structure zones in microemulsions. Nature, 1981, 293 (5834): 636-638.

[69] Yan F, Texter J. Surfactant ionic liquid-based microemulsions for polymerization. Chemical communications (Cambridge, England), 2006 (25): 2696-2698.

[70] Dasgupta A, Das D, Mitra R, Das P. Surfactant tail length-dependent lipase activity profile in cationic water-in-oil microemulsions. Journal of Colloid and Interface Science, 2005, 289 (2): 566-573.

[71] 朱文庆，许磊，马瑾，任建梅，陈亚芍．粒径可控纳米 CeO_2 的微乳液法合成．物理化学学报，2010,(5): 1284-1290.

[72] 戈磊，张宪华．微乳液法合成新型可见光催化剂 $BiVO_4$ 及光催化性能研究．无机材料学报，2009,(3): 453-456.

[73] 孙翼飞，巩宗强，苏振成，王晓光，图影．应用表面活性剂-生物柴油微乳液去除污染土壤中多环芳烃．环境工程学报，2012 (6): 2023-2028.

[74] Ueno K, Tokuda H, Watanabe M. Ionicity in ionic liquids: Correlation with ionic structure and physicochemical properties. Physical Chemistry Chemical Physics, 2010, 12 (8): 1649-1658.

[75] Blanchard L A, Brennecke J F. Recovery of organic products from ionic liquids using supercritical carbon dioxide. Industrial and Engineering Chemistry Research, 2001, 40 (1): 287-292.

[76] Blanchard L A, Hancu D, Beckman E J, Brennecke J F. Green processing using ionic liquids and

CO_2. Nature, 1999, 399 (6731): 28-29.

[77] Lateef H, Grimes S, Kewcharoenwong P, Feinberg B. Separation and recovery of cellulose and lignin using ionic liquids: a process for recovery from paper-based waste. Journal of Chemical Technology and Biotechnology, 2009, 84 (12): 1818-1827.

[78] Ma J, Hong X. Application of ionic liquids in organic pollutants control. Journal of Environmental Management, 2012, 99: 104-109.

[79] Scurto A M, Aki S N V K, Brennecke JF. Carbon dioxide induced separation of ionic liquids and water. Chemical Communications, 2003 (5): 572-573.

[80] Visser A E, Swatloski R P, Rogers R D. pH-dependent partitioning in room temperature ionic liquids: provides a link to traditional solvent extraction behavior. Green Chemistry, 2000, 2 (1): 1-4.

[81] Zhao D, Liu R, Wang J, Liu B. Photochemical oxidation-ionic liquid extraction coupling technique in deep desulphurization of light oil. Energy and Fuels, 2008, 22 (2): 1100-1103.

[82] Li C P, Xin B P, Xu W G, Zhang Q. Study on the extraction of dyes into a room-temperature ionic liquid and their mechanisms. Journal of Chemical Technology and Biotechnology, 2007, 82 (2): 196-204.